# SCIENCE AFTER BABEL

# SCIENCE AFTER BABEL

DAVID BERLINSKI

SEATTLE    DISCOVERY INSTITUTE PRESS    2023

## Description

Polymath and raconteur David Berlinski is at it again, challenging the shibboleths of contemporary science with his inimitable blend of deep learning, close reasoning, and rapier wit. In *Science After Babel* he reflects on everything from Newton, Einstein, and Gödel to catastrophe theory, information theory, and the morass that is modern Darwinism. The scientific enterprise is unarguably impressive, but it shows no sign of reaching the empyrean heights it seemed to promise a century ago. "It resembles Bruegel's Tower of Babel," Berlinski says, "and if it suggests anything at all, it suggests that its original plans have somehow been lost." Science endures. Scientism, it would seem, is guttering out.

## Library Cataloging Data

*Science After Babel* by David Berlinski

Library of Congress Control Number: 2023936544

ISBN: 978-1-63712-026-2 (Paperback), 978-1-63712-027-9 (EPUB), 978-1-63712-028-6 (Kindle)

BISAC: SCI080000 SCIENCE / Essays

BISAC: SCI034000 SCIENCE / History

BISAC: SCI027000 SCIENCE / Life Sciences / Evolution

## Publisher Information

Discovery Institute Press, 208 Columbia Street, Seattle, WA 98104

Internet: http://www.discoveryinstitutepress.com/

Published in the United States of America on acid-free paper.

First Edition, June, 2023.

# ADVANCE PRAISE

Many will read this book for the close, elegant reasoning, the astonishing erudition, or the mordant analysis. I confess I read it for the prose. "Vast sections of our experience might be so very rich in information"—I quote here from the discussion of our limited ability to define complexity— "that they stay forever outside the scope of theory and remain simply what they are: unique, ineffable, insubsumable, irreducible." See what I mean? Nobody but David Berlinski has ever employed such sweet, gorgeous prose in writing about science.

—**Peter Robinson**, Murdoch Distinguished Policy Fellow at the Hoover Institution and former speechwriter to President Ronald Reagan

Whether deconstructing the latest theory of everything or dishing on scientists and mathematicians he has known, whatever David Berlinski writes is delightful and profitable to read!

—**Michael Behe**, Professor of Biological Sciences, Lehigh University, author of *Darwin's Black Box*, *The Edge of Evolution*, and *Darwin Devolves*

If I were picking two books to be required reading for every college student in the United States, *Science After Babel* would be one. A striking and beautiful *and* absolutely necessary book. David Berlinski at his spectacular best.

—**David Gelernter**, Professor of Computer Science, Yale University

*Science After Babel* is a literary triumph. In it, David Berlinski masterfully exposes the hubris of scientific pretensions with a wit that dances deftly between the lines, unveiling profound insights with a refreshing candor. This book testifies to the author's penetrating intellect, inviting readers to reconsider the limits of scientific authority and reject facile invocations of science that demand assent at the expense of compelling evidence and rigorous thought.

—**William Dembski**, mathematician, philosopher, and former head of the Michael Polanyi Center at Baylor University; author of multiple groundbreaking works on the theory of intelligent design, including *The Design Inference* (Cambridge University Press, 1998)

Berlinski speaks wittingly as an insider to the sciences and their recent history. As a historian-philosopher of science, I recognize numerous valuable insights in this collection of arguments and memories. He captures the wonder of scientific inquiry without misplaced worship of speculative pronouncements made in its name. Berlinski is the most enjoyable antidote to scientism I know.

—**Michael Keas**, Lecturer in the History and Philosophy of Science, Biola University, author of *Unbelievable: 7 Myths About the History and Future of Science and Religion*

Dr. Berlinski explores everything from the complicated spawning behavior of salmon and the problems with the RNA World hypothesis to various acute challenges to modern evolutionary theory, including the Cambrian explosion, molecular machines, and the failure of punctuated equilibrium. As he shows, trouble is brewing for Darwin on other fronts as well—population genetics, taxonomy, behavioral psychology, and the philosophy of biology, to name just a few. In total, *Science After Babel* is a lively mix of deep scientific knowledge, literary skill, and humor. The work reveals why scientism's contemporary tower of babel has failed to reach the heavens. I highly recommend the book and hope it is widely read.

—**Ola Hössjer**, Professor of Mathematical Statistics, Stockholm University

David Berlinski is world class at combining devastating wit with even more devastating substantive criticism, and this book would be well worth its purchase price for the fine polemic essays it contains. But what makes it really special is the inclusion of hard-to-find historical and technical essays on mathematics, biology, and philosophy. In his popular writing, Berlinski has generally refrained from delving into all the gory details of his arguments, though we can sense a mastery of those details. Here the mastery is on unrestrained display, and the details can be fully enjoyed.

—**Stephen McKeown**, Assistant Professor of Mathematics, University of Texas at Dallas

In *Science After Babel* David Berlinski takes critical scholarly aim at many current day "scientific truths"—more properly shibboleths—including Darwinism, reductionism, the Standard Model of particle physics, and "talking" chimpanzees; and he shows how much nonsense often passes as secure scientific knowledge. Neo-Darwinism he describes as "empty," and in discussing the Standard Model he comments wryly, "Theories come and go." He also takes aim at a vast constellation of recent authors, including cosmologists Brian Greene and Lawrence Krauss, biologist Stephen Jay Gould, and philosopher of biology Michael Ruse.

The book is a delightful read delivered with great wit and erudition. We are treated to unique recollections—of his drinking coffee in Paris with René Thom, the founder of catastrophe theory; of the insane driving, also in Paris, of his friend, the mathematician, polymath, and leading French anti-Darwinist Marcel Schützenberger; and of a conversation with Noam Chomsky. Altogether the book represents an extraordinary and absolutely fascinating tour de force touching on topics as diverse as medieval Islamic astronomy and the great twentieth-century mathematician John von Neumann's reflections on the role of chance in evolution. The text is interspersed throughout with some beautiful descriptive writing—Mount Rainer's snow glimpsed flying out of SeaTac was "silent, sweeping, silvery, still, serene."

The book is a stunning intellectual achievement. Few authors could have written such a far-reaching, in-depth critique of so many current philosophical and scientific beliefs. *Science After Babel* is mandatory reading for anyone interested in a critical assessment of much current scientific thinking. No other recent publication comes close, and unquestionably this brilliant book establishes David Berlinski as one of the leading intellectuals of our time.

—**Michael Denton**, PhD, MD, former Senior Research Fellow in the Biochemistry Department at the University of Otago in Dunedin, New Zealand, author of *Evolution: A Theory in Crisis*, *Nature's Destiny*, and *The Miracle of Man*

# DEDICATION

*À*

*Stephen Meyer*

*De la part de son vieux camarade et ami.*

# CONTENTS

# INTRODUCTION

THE SCIENTIFIC REVOLUTION BEGAN IN THE SIXTEENTH CENTURY, and it began in Europe. No one knows why it happened nor why it happened where it happened, but when it happened, everything changed.

Until the day before yesterday, the imperial architects of the scientific revolution were well satisfied and sleek as seals. An immense tower was going up before their very eyes. The physicists imagined that shortly it would reach the sky; the biologists were satisfied that it had left the ground; and only the theologians were heard to observe that it would soon collapse.

The Tower is still there. It is, in fact, larger than ever. But it has neither reached the sky nor left the ground. It resembles Bruegel's Tower of Babel far more than the Chrysler Building, and if it suggests anything at all, it suggests that its original plans have somehow been lost. Some parts of the Tower are sound and sturdy; but, my goodness, the balustrade devoted to the multiverse—what were they thinking?

Who knows? In looking at the Tower, if we are moved to admire its size, we are also bound to acknowledge its faults. The algorithm and the calculus are the two great ideas of the scientific revolution. They are radically different. Algorithms belong to the world of things. The creation of numbers, Thierry of Chartres observed, *was* the creation of things. In the theory of recursive functions, some part of *thinginess* has been brought under rational control.

It is the continuum, on the other hand, that is essential to the calculus. If an algorithm is a part of the world of things, in war, Lewis Fry Richardson once remarked, *thinginess* fails. In quantum field theory, too. A quantum field is not a thing. The true continuum, René Thom once remarked to me, has no points: it reflects at a distance Freud's oceanic feeling—what Meister Eckhardt described as pure formlessness. And these, too, are ideas deep in human experience. In the calculus, and mathematical analysis generally, some part of the continuum has been brought under rational control.

Mathematical analysis and the theory of recursive functions are great achievements, but they are different; they answer to different imperatives; they are the work of different architects.

No wonder the Tower looks as it does. It is a miracle that it remains standing.

The result has been a popular culture littered with ideological detritus: atheism, of course, or naturalism, or materialism, or physicalism, or scientism, or even, God help us, trans-humanism. These are not very precise terms, nor do they denote very precise ideas. Naturalists can rarely say of naturalism anything beyond that it is natural.

"I come from a scientific background," David Chalmers modestly remarked. "I want everything to be natural," he added at once, "reduced to the simplest possible set of laws and entities."[1]

On this view, it is hard to see why *stuff happens* should not be considered a foundation for belief, the declaration requiring only two words and one substance.

Materialism has just a bit more by way of oomph. From a material base, as Marxists might say ominously, *everything*. Within contemporary physics, the deduction of everything from something is by no means complete and remains in that empyrean of assurances of which *your check is in the mail* is a notable example. Nor is the requisite something persuasively a material object. On current physical theories, that material base is occupied by various quantum fields, where, like so many electric eels, they occupy themselves in quivering with energy. Leptons and bosons emerge as field excitations, and so does everything else.

The great merit of materialism has always been its apparent sobriety. A world of matter? Look around! Bang the table, if necessary. Quantum fields do not encourage a look-around. There is no banging them beyond banging on about them. And for the most obvious of reasons. "Quantum field theory," Lisa Randall writes, "the tool with which we study particles, is based upon eternal, omnipresent objects that can create and destroy those particles."[2]

This is an account that suggests the dominion of Vishnu as much as metaphysical materialism, a point not lost on Indian physicists. And it may well change, that account, those infernal quantum fields vanishing tomorrow in favor of otherwise unexpected entelechies.

There remains the curious fact that no one much likes what everyone accepts. What everyone accepts is something like the scientific system of belief. It is to this system that every knee must bend, with *trust the science* functioning both as an inducement and an admonition. If contemporary scientists are not voyaging strange seas *alone*, to recall Wordsworth's epitaph for Isaac Newton, they are yet determined to put as much distance as possible between themselves and dry land. That quantum mechanics makes no sense is widely celebrated as one of its virtues. Not a day passes in which its weirdness is not extolled. As much might be said of the Eucharist, but with this considerable difference: scientific weirdness tends inexorably toward a kind of bleakness. *"Le silence éternel de ces espaces infinis m'effraie,"* Pascal remarked[3]; and had he been acquainted with contemporary cosmologies in which the universe is destined to gutter out into something barren, formless, flaccid, lightless, and large, his anxieties may well have been proportionally increased.

The scientific system of belief remains what it was: implacable and unavoidable. There is no getting around it and so no getting out of it. The notes, incidental remarks, essays, and reviews that comprise this book represent an inside job, and it is in the nature of inside jobs that the inside jobber cannot expect outside help. It is an irony of any imperial enterprise, whether political, social, or intellectual, that it determines the conditions under which it may be criticized.

For this reason, what I have written in this book is an exercise in *self-criticism* as much as anything else. I often wish that things were otherwise. "The Shepherd in Virgil grew at last acquainted with Love, and found him a Native of the Rocks."[4] No one quite gets what he wants—not in life, nor in love, nor, as it happens, in writing critical essays.

**Paris, 2023**

# I. Darwin, Checking In

# 1. Haunting History

D ARWIN COMPLETED HIS MASTERPIECE, *ON THE ORIGIN OF SPECIES*, in 1859. He was then forty-nine, ten years younger than the century, and not a man inclined to hasty publication. In the early 1830s, he had journeyed around the globe as a naturalist aboard HMS *The Beagle*. The stunning diversity of plant and animal life that he saw impressed him deeply. Prevailing biological thought held that each species is somehow fixed and unalterable: looking backward in time along a line of dogs, it is dogs all the way. Five years at sea suggested otherwise to Darwin.

By 1837, Darwin realized that the evidence of evolving species he had witnessed on the voyage might hold for the rest of life and this, in turn, suggested the dramatic hypothesis that, far from being fixed and frozen, the species that now swarm over the surface of the earth evolved from species that had come before in a continuous, phylogenetic, saxophone-like slide.

What Darwin lacked in 1837 was a theory to account for speciation. The great ideas of fitness and natural selection evidently came to him before 1842, for by 1843 he had prepared a version of his vision and committed it to print in the event of his death. He then sat on his results in an immensely slow, self-satisfied, thoroughly constipated way until news reached him that Alfred Russel Wallace was about to make known *his* theory of evolution. Wallace collected data in the East Indies, and considering the same problem that had earlier vexed Darwin, hit on precisely Darwin's explanation. The idea that Wallace might hog the glory was too much for the melancholic Darwin: he lumbered into print just months ahead of his rival; but in science, as elsewhere, even seconds count.

The theory that Darwin proposed to account for biological change is a conceptual mechanism of only three parts. It involves, in the first instance, the observation that small but significant variations occur naturally among members of a common species. Every dog, for example, is doggish in his own way. Some are fast, others slow, some charming, others suitable only for crime. Yet each dog is essentially dog-like to the bone, a dog, *malheureusement*, and not some other creature. Darwin wrote before the mechanism of genetic transmission was understood, but he inclined to the view that variations in the plant and animal kingdoms arise by *chance* and are then passed downward from fathers to sons.

The biological world, Darwin observed, striking now for the second point to his three-part explanation, is arranged so that what is needed for survival is generally in short supply: food, water, space, tenure. Competition thus ensues, with every living creature scrambling to get his share of things and keep it. The struggle for life favors those organisms whose variations give them a competitive edge. Such is the notion of *fitness*. Speed makes for fitness among the rabbits, even as a feathery layer of oiled down makes the Siberian swan a fitter fowl. At any time, those creatures fitter than others will be more likely to survive and reproduce. The winnowing in life effected by competition Darwin termed *natural selection*.

Working backward, Darwin argued that present forms of life, various and wonderful as they are, arose from common ancestors; working forward, that biological change, the transformation of one species to another, is the result of small increments that accumulate across the generations. The Darwinian mechanism is both random and determinate. Variations occur without plan or purpose—the luck of the draw—but Nature, like the House, is aggressive, organized to cash in on the odds.

# 2. Misprints in the Book of Life

Everything that lives, lives but once. To pass from fathers to sons is to pass from a copy to a copy. This is not quite immortality, but it counts for something, as every parent knows. The higher organisms reproduce themselves sexually, of course, and every copy is copied from a double template. Bacteria manage the matter alone, and so do the cells within a complex organism, which often continue to grow and reproduce after their host has perished, unaware for a brief time of the gloomy catastrophe taking place around them.

It is possible, I suppose, that each bacterial cell contains a tiny copy of itself, with the copy carrying yet another copy. Biologists of the early eighteenth century, irritated and baffled by the mystery of it all, thought of reproduction in these terms. Peering into crude, brass-rimmed microscopes, they persuaded themselves that on the thin-stained glass they actually saw an homunculus. The more diligent of the biologists then proceeded to sketch what they seemed to see. The theory that emerged had the great virtue of being intellectually repugnant.

Much more likely, at least on the grounds of reasonableness and common sense, is the idea that the bacterial cell contains what Erwin Schrödinger called a *code script*—a sort of cellular secretary organizing and recording the gross and microscopic features of the cell. Such a code script would logically be bound to double duty. As the cell divides in two, it, too, would have to divide without remainder, doubling itself to accommodate two cells where formerly there was only one. Divided, and thus doubled without loss, the code script would require powers sufficient to organize anew the whole of each cell.

The code script that Schrödinger anticipated in his moving and remarkable book, *What is Life?*, turns out to be, in significant measure, DNA, a long and sinewy molecule shaped rather like a double-stranded spiral. The strands themselves are made of stiff sugars, and stuck in the sugars, like beads in a sticky string, are certain chemical bases: adenine, cytosine, guanine, and thymine—A, C, G, and T, in the now universal abbreviation of biochemists. It is the varied alternation of these bases along the backbone of DNA that allows the molecule to store information.

One bacterial cell splits in two. Each is a copy of the first. All that physically passes from cell to cell is a strand of DNA. The message that each generation sends faithfully into the future is impalpable, abstract almost, a kind of hidden hum against the coarse wet plops of reproduction, gestation, and birth itself. James Watson and Francis Crick provided the correct description of the chemical structure of DNA in 1952.

They knew, as everyone did, that somehow the bacterial cell, in replicating itself, sends messages to each of its immediate descendants. They did not know how. But the chemical structure of DNA, once elaborated, suggests irresistibly a mechanism for both self-replication and the transmission of information. In the cell itself, strands of DNA are woven around each other, and, by an ingenious twist of biochemistry, matched antagonistically: A with T, and C with G. At reproduction, the cell splits the double strand of DNA. Each half floats for a time, a gently waving genetic filament; chemical bonds are then repaired as the bases fasten to a new antagonist, one picked from the ambient broth of the cell. The process complete, there are now two strands of double-stranded DNA where before there was only one.

What this account does not provide is a description of the machinery by which the new cells are actually organized. To the molecular biologist, the cell appears as a small sac enclosing an actively throbbing biochemical machine. What the machine extrudes are long and complex molecules constructed from a stock of twenty amino acids.[1] Such are the *proteins*. The order and composition of the amino acids along a given chain determine which protein is which. The cell contains a complete record of the right proteins, as well as the instructions required to assemble them directly. The sense of genetic identity that marks *E. coli* as *E. coli*, and not some other bug, is thus *expressed* in the amino acids by means of information *stored* in the nucleotides.

The four nucleotides, we now know, are grouped together in a triplet code of sixty-four codons, or operating units. A particular codon is composed of three nucleotides. The amino acids are matched to the codons: C-G-A, for example, to arginine. In the translation of genetic information from DNA to the proteins, the linear ordering of the codons themselves induces a corresponding linear ordering first onto an intermediary, messenger RNA, and then onto the amino acids themselves—this via yet another intermediary, transfer RNA. The sequential arrangement of the amino acids influences the chemical configuration of a protein. Molecular biologists often allude to the steps so described as the *central dogma*, a queer choice of words for a science.

The Austrian monk Gregor Mendel founded the science of genetics on purely a theoretical notion of the gene. In DNA one looks on genetics bare: the ultimate unit of genetic information is the nucleotide. All that makes for difference, and hence for drama, in the natural world, and that is not the product of culture, art, artifice, accident, or hard work, all this, which is brilliantly expressed in perishable flesh, is a matter of an ordering of four biochemical letters along two ropy strands of a single complex molecule.

The central dogma describes genetic replication, but the concepts that it scouts illuminate Darwinian theory from within. Whether as the result of radiation or chemical accident, letters in the genetic code may be scrambled, with one letter shifted for another; entire codons may be replaced, deleted, or altered—*mutations* in the genetic message. They are arbitrary, because they are unpredictable, and yet enduring, because they are genetic. The theory by which Darwin proposed to account for the origin of species and the nature of biological diversity now admits of expression in a single English sentence: Evolution, or biological change, so the revised, the *neo*-Darwinian theory runs, is the result of natural selection working on random genetic mutations.

# 3. A System of Belief

THE THESIS THAT THEORETICAL BIOLOGY CONSTITUTES A KIND OF intellectual Uganda owes much to the theory that biology is itself a *derivative* science, an analogue to automotive engineering or dairy management, and, in any case, devoid of those special principles that lend to the physical or chemical sciences their striking mahogany luster. Naïve physicists—the only kind—are all too happy to hear that among the sciences physics occupies a position of prominence denied, say, to horticulture or agronomy. The result is *reductionism from the top down*, a crude but still violently vigorous flower in the philosophy of science. The physicist or philosopher, with his eye fixed on the primacy of physics, thus needs to sense in the other sciences—sociology, neurophysiology, macramé, whatever—intimations of physics, however faint. This is easy enough in the case of biochemistry. Chemistry is physics once removed; biochemistry, physics at a double distance. Doing biochemistry, the theoretician is applying merely the principles of chemistry to living systems. His is a reflected radiance.

In 1831, the German chemist Friedrich Wöhler synthesized urea, purely an organic compound—the chief ingredient in urine, actually—from a handful of chemicals that he took from his stock and a revolting mixture of dried horse blood. It was thus that organic chemistry was created, an inauspicious beginning, but important nonetheless, if only because so many European chemists were convinced that the attempt to synthesize an organic compound would end inevitably in failure.

The daring idea that *all* of life—I am quoting from James Watson's textbook, *The Molecular Biology of the Gene*—will ultimately be understood in terms of the "coordinative interaction of large and small molecules" is now a commonplace among molecular biologists, a fixed point in the wandering system of their theories and beliefs. The contrary thesis that living creatures go quite beyond the reach of chemistry biochemists regard with the alarmed contempt that they reserve for ideas they are prepared to dismiss but not discuss. Francis Crick, for example, devotes fully a third of his little monograph, *Of Molecules and Men*, to a denunciation of vitalism almost ecclesiastical in its forthrightness and utter lack of detail. Like other men, molecular biologists evidently derive some satisfaction from imagining that the orthodoxy they espouse is ceaselessly under attack.

Curiously enough, while molecular genetics provides an interpretation for certain Darwinian concepts—those differences between organisms that Darwin observed but could not explain—the Darwinian theory itself resists reformulation in terms either of chemistry or physics. This is a point apt to engender controversy. Analytic philosophers cast reduction as a logical relationship. Given two theories, the first may be reduced to the second when the first may be *derived* from the second. The standard and, indeed, the sole example of reduction successfully achieved involves the derivation of thermodynamics from statistical mechanics. In recent years, philosophers have come to regard the concept of direct reduction with some unhappiness. There are problems in the interpretation of historical terms—the Newtonian concept of mass, for example—and theories that once seemed cut from the same cloth now appear alarmingly incommensurable.

To speak of the *logical* structure of biological thought is at least tentatively to suggest that biological theories have something even vaguely discernible *as* a logical structure. This represents, I think, the lingering influence on the philosophy of biology of standards current in the philosophy of physics. Now theories in physics quite often are logically disorganized and, despite the animadversions of philosophers, none the worse for that. Their intellectual robustness is, I think, a function of the fact that physical objects are almost always entirely determined by physical theories. To the philosopher who wishes to know what a quark *is*, the physicist need only point to the laws that describe quarks—the principles of quantum chromodynamics, say. The philosopher who insists that this is all very well and demands to know further what a quark *really* is, has asked an unanswerable question. In the case of life, however, the objects scouted by biological theory have an antecedent conceptual existence, one that is quite indifferent to expression in theoretical terms. These objects are fixed in our imagination by their position within a dense matrix of concepts, with the matrix itself animated by dim, inarticulate biological throbbings.

In comparison to physical theories, biological theories are circumscribed; the philosopher asking innocently for an account of life is hardly in a position to dismiss on principled grounds any number of possible answers—a play of biochemical forces, physics in its most complex state, the coordinative interaction of large and small molecules (Watson's answer), aspects of the Mind of God, the structures forged to protect the gene, the appearance in the universe of pity and terror. It is some measure of the confusion in contemporary thought that each of these answers seems roughly right.

But no matter the degree to which molecular biology is logically disorganized, the definition of reduction that I have cited is incomplete as well as irrelevant. In Mendelian genetics, the concept of a gene is theoretical, and genes figure in that theory as abstract entities. To what should they be pegged in molecular genetics in order to reduce the first theory to the second? DNA, quite plainly, but how much of the stuff counts as a gene? "Just enough to act as a unit of function," argues Michael Ruse, a philosopher whose commitment to prevailing orthodoxy is a model of steadfastness. But in biochemistry the notion of a unit of function is otiose, unneeded elsewhere. To the extent that biochemistry is molecular genetics, it does not reflect completely Mendelian genetics; to the extent that it does, it is not biochemistry, but biochemistry beefed up by a conceptual padded shoulder.

What holds in a limited way for molecular genetics holds in a much larger way for molecular biology. Concepts such as *code* and *codon, information, complexity, replication, self-organization, regulation,* and *control*—the items required to make molecular biology *work*—are scarcely biochemical. The biochemist following some placid metabolic pathway need never appeal to them.

Population genetics, to pursue the argument outward toward increasing generality, is a refined and abstract version of Darwin's theory of natural selection, one that is applied directly to an imaginary population of genes. Selection pressures act on the molecules themselves, a high wind that cuts through the flesh of life to reach its buzzing core. Has one here achieved anything like a reduction of Darwinian thought to theories that are essentially biochemical, or even vaguely physical? Hardly. The usual Darwinian concepts of fitness and selection remain unvaryingly in place. These are ideas, it goes without saying, that do not figure in standard accounts of biochemistry, which very sensibly treat of valences and bonding angles, enzymes, fats, and polymers—anything but fitness and natural selection.

To the standard conditions on reduction, then, I would add a caveat: no reduction by means of inflation. The Darwinian theory of evolution is the great, global, organizing principle of biology, however much molecular biologists may occupy themselves locally in determining nucleotide sequences, synthesizing enzymes, or cloning frogs. Those biologists who look forward to the withering away of biology in favor of biochemistry and then physics are inevitably neo-Darwinians, and the fact that this theory—*their* theory—is impervious to reduction they count as an innocent inconsistency.

Theoretical biologists still cast their limpid and untroubled gaze over a world organized in its largest aspects by Darwinian concepts. But unlike the theory of relativity, which Einstein introduced to a baffled and uncomprehending world in 1905, the Darwinian theory of evolution has never quite achieved canonical status in contemporary thought, however much its influence may have been felt in economics, sociology, or political science. If mathematical physics offers a vision of reality at its most comprehensive, the Darwinian theory of evolution, like psychoanalysis,

Marxism, or the Catholic faith, constitutes, instead, a system of belief. Like Hell itself, which is said to be protected by walls that are seven miles thick, each such system looks especially sturdy from the inside. Standing at dead center, most people have considerable difficulty in imagining that an outside exists at all.

# 4. THE EVIDENCE
# FOR EVOLUTION

EVOLUTION, IT IS HELD, TAKES PLACE OVER A VERY LONG TIME. THE human hand evolved from the inhuman paw, step by step, one incremental improvement following another, a tortuous process, endlessly delayed, endlessly extended. No one, of course, has actually seen the whole business at work. British biologists in the north of England are said to have observed the peppered moth species changing its wing coloring in order to maintain its mimetic protection. Laboratory insects, most notably the fruit fly *Drosophilia*, have been tracked through a series of evolutionary changes in wing structure and color. But these anecdotal examples provide no real evidence that Darwinian mechanisms are at work and no evidence, surely, that Darwinian theories *explain* the process by which a new species arises from one that is old.[1]

On this matter, the science of paleontology has some bearing. Long periods of geological upheaval during the Earth's early history have had the effect of trapping a variety of flora and fauna beneath the shifting tides of geological debris and freezing them there in an easily visible pattern. The record of fossils so laid down makes up a *paleolithic stratum*; the various strata are read from the bottom up, and by moving upward, paleontologists can form a picture of the progression of organisms over time.

One might expect that the record in the rocks and the neo-Darwinian theory would fit, hand to glove, with the bottom layer exhibiting the simplest microorganisms, and each succeeding layer trailing off continuously into the next. In fact, the fit between theory and data is poor. Virtually no multicellular fossils have been found in the Precambrian rock strata prior to the Ediacaran; and with the beginning of the Cambrian, one sees an explosive proliferation of bilaterian animal life forms—*de novo*, as it were, appearing abruptly, without obvious ancestors in the preceding strata. Here and elsewhere there are great gaps in the fossil record, strange discontinuities, an almost complete absence of intermediate forms, as if, to everyone's astonishment, the individual species had been sculpted separately, just as the medievals believed all along.

The paleontological record is, of course, hardly proof of special creation, but neither does it make for a great happiness among biologists. "Despite the bright promise that paleontology provides as a means for 'seeing' evolution," David Kitts has written in the journal *Evolution*, "it has presented some nasty difficulties for evolutionists, the most notorious of which is the presence of gaps in the fossil record. Evolution requires intermediate forms between species and paleontology does not provide them."[2]

Of course, if the fossil record does not fit the theory, it is always possible to adjust the theory to fit the record. In science, an enterprising theoretician has several degrees of freedom within which to maneuver before the referee reaches ten and the final bell comes to clang. Stephen Jay Gould, who was trained as a paleontologist, surveyed the fossil evidence early in the 1970s and came to the obvious conclusion that either the theory or the evidence must go. What went, on his scheme of things, was the neo-Darwinian orthodoxy by which species change into different species by means of an endless series of infinitesimal changes, continuously, like the flow of syrup.

Instead, Gould argued, biological change must have been discontinuous, with vast changes taking place too quickly for the fossil record to register the series of small steps from A to B. Such was his model of *punctuated equilibria*. It fits the fossil record far better (if it makes sense, even, to talk of scientific fit here), but the model achieves faithfulness to the facts only by chucking out the chief concepts of the Darwinian theory itself, and while paleontologists have been glad to have had Gould's company, evolutionary theorists have looked over what he has written with the cool, slitted, appraising glance of a butcher eyeing a sheep.

§

In an entirely different way, some philosophers have always found something fishy in the Darwinian theory of evolution. An obvious sticking point is the concept of fitness itself. If by the fitter organisms, biologists mean merely those that survive, then the doctrine that natural selection winnows out those organisms that are not fit expresses a triviality. This is a logical point and not a matter of experiment or research. The biologist who wishes to know why a species that represents nothing more than a persistent snore throughout the long night of evolution should suddenly or slowly develop a novel characteristic will learn from the neo-Darwinian theory only that those characteristics that survive, survive in virtue of their relative fitness. Those characteristics that are relatively fit, on the other hand, are relatively fit in virtue of the fact that they have survived. This is not an intellectual circle calculated to inspire confidence.

Biologists, it goes without saying, reject with florid indignation the idea that the Darwinian theory is empty. Ernst Mayr, for example, asserts that "to say that this is the essence of Darwin's reasoning is nonsense." These are forthright words. Yet in writing of traits that have *no* apparent selective advantage, Mayr argues that "the mere fact that such traits have become established makes it highly probable that they are the result of selection."[3] If this does not mean simply that those traits that survived survived, whatever else it might mean is very obscure indeed.

The evaporation of content from the notion of fitness is nowhere more apparent than in mathematical genetics, a subject with a reputation for pointlessness inferior only to that of modern tax law. In Albert Jacquard's *The Genetic Structure of Populations*, the crude idea of fitness has disappeared in favor of the notion of selective value, which Jacquard defines as a number "proportional to the probability of survival from conception to adulthood of individuals of [a given] phenotype."[4] Whatever it is that accounts for the probability of survival, however, has become impalpable, a power registered only through its effects. This is rather as if an engineer were to define *stress* as a number proportional to the probability of failure, without ever inquiring why a particular beam buckled, or why beams buckle in general, or whether stress and structural failure represent two concepts or only one.

The doctrine that survival favors the survivors is what logicians call a *tautology*, a statement that is all form and no content. For obvious reasons, evolutionary biologists are uncomfortable with the idea that the chief claim of their theory is roughly on the intellectual order of the declaration that whatever will be, will be. Every organism, to continue the argument, needs to accomplish certain biological tasks simply in order to stay alive. The antelope must get food in order to keep on getting food, and it must avoid becoming food for those animals, such as the lion, that nourish themselves by eating other animals. In looking for food, there is an obvious advantage to claws, teeth, speed, keen sight, or a first-name familiarity with the headwaiter. Is this not, then, a step in the right— the Darwinian—direction? I am dubious. The obvious cases do appear to sort themselves out according to Darwinian principles. But were the pig to be born with wheels mounted on ball bearings instead of trotters, would it be better off on some scale of porcine fitness? No one knows, although some guesses are possible. The general idea that a biological organism requires certain powers if it is to survive is inadequate to explain the *specific* form those powers happen to take.

The petals of the orchid *Ophrys apifera*, for example, exhibit a design similar to the design inscribed on the genitalia of the female bee. The mimetic effect is quite striking, at least to the bee. But why such a complex mechanism, such fantastic drollery, when no other orchid requires anything like it?

The Pacific salmon, to take another example—the last—is a magnificent fish—long, heavy, silvery, sleek. It is spawned, among other places, in the streams that ultimately feed into Puget Sound and then wash out to sea. After a number of years spent in the open ocean, the salmon returns to its spawning beds to lay eggs and die.

In the late fall, as the thin sun casts wicked-looking cloud-stained streaks over the northern lowlands, thousands upon thousands of salmon begin to course through the inland sea, seeking out the mouth of the particular river that will ultimately wind backward through the evergreen forests to the very stream in which they were spawned. The geographic precision and unerring instinct that enables these fish to find their spawning grounds is largely a mystery. Where there is a means, there is a mechanism, and, presumably, some set of features enables the salmon to head home again—distinctive smells, perhaps, or subtle magnetic signals, or the tang and sheen of the air.

The journey undertaken by the Pacific salmon is unbearably poignant because it is at once desperate and directed toward death. This is to describe things in the abstract. To stand alongside the Duwamish or Steilacoom rivers in mid-October, and watch the headwaters course with salmon, is to be struck by the unbearable grandeur of life itself.

For the salmon, going backward along a river means going upward: over waterfalls, past rock formations, swimming all the time against the current and toward an ever-narrowing wedge of water. By the time the fish have pushed several miles upstream, they are exhausted. Their skins are grey and mottled with parasitic fungi. Their eyes are no longer clear. They expend energy in spurts, panting heavily when they come to still waters. Yet they are still frenzied, maddened by some vibrant call. Along the banks of the river, eagles sit patiently on the drooping branches, hardly bothering to fish. What they need will wash to shore. In the end, only a handful of fish make it to the headwaters of the rivers, and there they lay their eggs. But a handful is all that is needed, for each female deposits thousands of eggs, and of these thousands, thousands yet again will live to complete the cycle, to swim in the great northwest rivers, flourish for a while in the open ocean, and then swim up the rivers to lay their eggs and die.

I myself have not seen any of this, of course, but I am sure that it takes place.

Why do they bother, these salmon? No other fish requires such an elaborate apparatus of misery and migration. Perhaps the Pacific salmon was originally a freshwater fish separated somehow from its spawning grounds; perhaps they require fresh water to spawn. Perhaps the cycle is dictated by other factors. Yet here is the point that should provoke a doubt. The desirability, speaking fishwise, of an infinitely complicated reproductive routine is never demonstrated within Darwinian theory nor derived from general qualitative principles.

§

I pass now to an account of certain modest triumphs of evolutionary thought. The subject at hand is monkey testicles; the question at large, why some of them are more substantial than others. Harcourt, Harvey, Larson, and Short, a quartet suggesting a firm of accountants, studying precisely this issue, reasoned that in polygamous primate systems in which each male competes with every other male for mating rights, those males with the largest, heaviest-hanging testicles should inevitably be favored in the competition. They do not say why. It follows that the ratio of testicle weight to body size should be quite different in chimpanzee societies, which are promiscuous, and gorilla societies, which are not.

With careful restraint the authors note that "they have tested the hypothesis across a wide range of primates," although, for understandable reasons, they provide no data concerning the actual protocols of measurement. Their work has been discussed by Martin and May, who observe that the *general* relationship between polygamy and testicular weight, what Spanish-speaking researchers might call *cojonismo*, fails in the case of the polygamous squirrel monkey, whose breeding season is short but whose testicles are small when compared to the steadfastly monogamous cotton-top tamarin.

Having explored such facts in a human context only furtively, from the corner of a wet eye, so to speak, I cannot comment on the reliability of the research, but I am all for science, wherever it may lead—so long, of course, as I do not have to follow.

# II. Darwin, Checking Out

# 5. PARTIR C'EST
# MOURIR UN PEU

E VERY JOURNEY, I SUPPOSE, IS AN ESCAPE. I LEFT FOR PARIS ON A dark northwest day in the fall of the year. Everything was dripping—the spiky evergreen leaves, the ends of staircase banisters, the trailing edge of billboards; even my breath entered the unmoving air as a humid puff and appeared for a moment to form a cloud. I clambered into the airport bus, bumping my head as I did, my suitcase banging irresolutely on the narrow stairs behind me. The black rubber mounting to my window formed a perfect frame: in the background, where Giotto would have painted a sunny mythical mountain, there was an unbroken ridge of wet green evergreens; in the middle, where Memling might have painted a turbid square, a dismal cluster of suburban shops—*Bruno's Pizza, Toys for Tots*, a restaurant inaccurately calling itself *The French Connection*, a massage parlor specializing in late-night commuters with fancy cravings; and in the near foreground, perfectly composed, a figure in a blue peacoat, a face that I knew as well as I knew my own, short black hair, tense eyes, the chin tilted just lightly, two mascara-streaked tears wandering over the cheeks.

The plane rumbled and groaned and clanked down the long runway at SeaTac and lumbered into the air. Below I could see the heavy cold waters of Puget Sound and the dense somber stands of spruce on Vashon Island. As the plane banked, and then turned steeply toward the north and east, it broke through the low-lying layer of cumulonimbus clouds that covers the northwest for much of the year, and there, impossibly bright, utterly clear, was Mount Rainier, sunlight on its snows, silent, sweeping, silvery, still, serene.

Some years before, I had spoken to Noam Chomsky about the theory of evolution. He told me to see Tom Bever at Columbia University, and then, almost as an afterthought, added that the man I really ought to talk to was Marcel Schützenberger in Paris. Chomsky told me that while a Junior Fellow at Harvard, Bever had driven James Watson mad with his questions about the theory of evolution. It was an image I cherished until I actually met Bever.

I thought little about Schützenberger until I published my first book; I sent him a copy with a hopeful note. Within days I received a handwritten response, thanking me for the book; days later, I received a second note. My little book, Schützenberger wrote, was a "masterpiece of critical thinking." I had never been accorded praise on this order. I was deeply impressed.

My purpose now in traveling to Paris was ostensibly to discuss the theory of evolution with Schützenberger. We had agreed by cable that he would arrange a hotel for me, something cheap.

I arrived in Paris that afternoon, and after a bouncing taxicab ride in a decrepit Citroën smelling vilely of Gauloises, I discovered that *Le Grand Hôtel de Paris* (three stars gleaming on the door, *bien sûr*) was located just off the Rue Monge, a street of positively supreme seediness, and that the hotel itself was small, dark, ugly, cramped, and battered. My room was dank. The wallpaper on both walls had peeled back, the better to accommodate a troop of bedbugs, who, in single file, were marching from left to right, arising from one crack in the wall to disappear after they had crossed the ceiling into another. There was a chipped porcelain washstand in the room, but no shower, and no toilet either, except for the low-lying, dust-filled bidet. When I sank onto the narrow bed, the thing wheezed melodiously, the mattress promptly curving into a parabola.

The monotonous hum of the jet exhaust was still very much in my ears; I looked up at the ceiling and endeavored to bunch up the curious tubular pillow that one finds in French hotels. I drifted off and actually slept for a few moments until roused unceremoniously by the hotel's clerk, who had, just an hour before, handled my American passport with the air of a man examining a dead fish, and was now standing outside my room knocking rhythmically on the door. I had a telephone call. It was Schützenberger.

"You come to lunch, yes?" he said, jumbling tenses and moods in a single sentence, with only the charming imperative left intact. I said that I would.

Schützenberger lived in the sixteenth *arrondissement,* in the sort of nondescript flat that in Paris signifies status but not necessarily wealth. The flat itself consisted of a series of large, sparsely furnished rooms; from somewhere, far away, the sound of chamber music drifted into the apartment. There were several grim pictures in the living room. In one, a stout, handsome woman stood staring somberly into space; in another, a small boy in shorts stood posed next to a drab horse, a brown, academic, nineteenth-century French landscape behind the two of them. The severe, iron-edged, double-doored windows were closed. It was very warm.

Schützenberger was somewhat shorter than medium height, very thin, in his late fifties, I judged, with thick, long, grey hair, which flopped over his forehead and which he continually swept back over his head. At the door, where he greeted me shyly, he carried himself with severe erectness, as if his lower spine had been fused, and when we walked to the living room, I noticed that like many Frenchmen he held his elbows close to his sides.

"It is a *fantastic* pleasure to see you," he said in English, but with an accent so strong and so melodious that I thought for a moment that he was speaking French.

We sat in large, old, comfortable chairs and talked. After a time, Schützenberger's Javanese wife, Hariati—a solid, immensely dignified, slow-moving woman with deep, wide, chocolate-brown eyes—came in with lunch.

When we had finished, we agreed to meet and talk again during the week. Our idea was to publish a paper together on the neo-Darwinian theory of evolution. When it was time for me to go, Schützenberger drove me back to my hotel. He was an insanely wretched driver, and once went several blocks out of his way in order to pass a driver who had passed him first. "Oh, the *fuckair*," he said with satisfaction after we had overtaken the man and cut him off.

# 6. Darwin and the Mathematicians

**Evolution News:** In the past, you've remarked about mathematicians and their opinions of Darwin's theory of evolution. They were skeptical, you said; *very* skeptical. John von Neumann was an example. How do you know that about him and about other mathematicians?

**David Berlinski (DB):** How do I know? Here's how: I have been close to a number of mathematicians, and friends with others: Daniel Gallin (who died before he could begin his career), M. P. Schützenberger (my great friend), René Thom (a friend as well), Gian-Carlo Rota (another friend), Lipman Bers (who taught me complex analysis and with whom I briefly shared a hospital room, he leaving as I was coming), Paul Halmos (a colleague in California), and Irving Segal (a friend by correspondence, embattled and distraught). Some of these men I admired very much, and all of them I liked.

I had many other friends in the international mathematical community. We exchanged views; I got around.

Among the mathematicians that I knew from very roughly 1970 to 1995, the general attitude toward Darwin's theory was one of skepticism. These days, I do not get around all that much, and whatever the mathematician's pulse, I do not have my finger on it. But the reactions of which I speak were hardly surprising. Until recently, mathematicians have been skeptical of any discipline beyond mathematics, and I say until recently because attitudes as well as times have changed.

In talking of the mathematician's skepticism, I mentioned von Neumann because his name was widely known. I might have mentioned Gian-Carlo Rota. He despised the enveloping air of worship associated with Darwin; he thought biology primitive and dishonest.

How do I know this? I know it because we were close friends, and because he said so. He said so to me.

Gian-Carlo had, in fact, read closely one of my unpublished papers on Darwinian evolution. Written in late 1970, during my stay at IIASA (the International Institute for Applied Systems Analysis), my essay later made its way into *Black Mischief: Language, Life, Logic & Luck*, stripped by then of almost all of its technical details.

A few mathematicians at IIASA had already read what I had written; they had, after all, encouraged me to write it. When we met later in the year, Gian-Carlo offered me his delighted agreement, which he extended in the spirit of *it's about time*; he urged me to keep at it; he considered publishing my essay in his journal; but after some back and forth between us, he decided that it would be best were I never to publish another word on the subject.

Gian-Carlo was a man of very refined political sensibilities.

It is a mistake to read back into the recent past the political and emotional structure of discussions now current.

Reading things backwards is vulgar as intellectual history and false to the facts—vulgar because it assigns an aspect of permanence to our own obsessions; and false because it distorts the play of forces playing just a few decades ago.

In the first part of the twentieth century, Darwin v. Dissent had not yet acquired its riveting incarnation as a melodrama of intolerance. No heresy, no heretics is a useful proverb, and using, say, 1950 as a reference point, there were no heretics among the mathematicians because there was yet no heresy. Darwin's theory was not then considered totemic; and his touch was not widely understood to cure erysipelas. Darwin v. Dissent is of our time and place.

Now von Neumann turned from pure mathematics in the 1940s and the early 1950s. Like so many other mathematicians and physicists, he regarded the theory of evolution as a placeholder, the full and so the real theory waiting somewhere in the wings of time.

When Erwin Schrödinger published *What is Life?* in 1944, he electrified the mathematicians and the physicists; and he influenced profoundly biologists such as Francis Crick, the latter a form of Rural Electrification, I suppose. Schrödinger's impact is easy to understand: he gave biologists a set of ideas that they had been unable to give themselves.

"How can the events in space and time," Schrödinger asked, "which take place within the spatial boundary of living organism be accounted for by physics and chemistry?"

These words were written in 1944.

"The preliminary answer which this little book will endeavor to expound," Schrödinger went on to say, "... can be summarized as follows: the *obvious inability* of present-day physics and chemistry to account for such events is no reason at all for doubting that they can be accounted for by those scientists."

I have placed in italics words that establish Schrödinger's attitude. I do not know which word to stress more: the obviousness of an inability, or the inability of an obviousness.

So I have stressed them together. They reflect the attitude of mathematicians and physicists toward biology during the 1940s, 1950s, and at least a part of the 1960s.

Having asked for a clue, Schrödinger found one on his own. He predicted the existence of a code script, one governing heredity. Just eight years later, Watson and Crick published the first of their two great papers on the structure of DNA.

Everyone took notice, the biologists because Schrödinger had been prophetic, and the mathematicians and physicists because Schrödinger had been one of their own.

These remarks belong to the considerable category of things that must be kept in mind.

So keep them in mind.

What was it that comprised von Neumann's skepticism about evolution?

It was an attitude in three aspects. Von Neumann, in the first place, saw what mathematicians had seen since Darwin first published his theory. The theory required life to clamber over some very sobering improbabilities; indeed, it seemed to require miracles. Other mathematicians were making the same point. "The formation within geological time of a human body," Kurt Gödel remarked in conversation with Hao Wang, "by the laws of physics (or any other laws of similar nature), starting from a random distribution of elementary particles and the field, is as unlikely as the separation by chance of the atmosphere into its components."

Note the word geological. Gödel, like everyone else, quite understood that Darwin's theory played out over long stretches of time. He might even have grasped the concept of natural selection, commonly said to be too difficult for all but a handful of initiates. He was skeptical nonetheless. It was precisely to do battle against this kind of skepticism that Richard Dawkins wrote *The Blind Watchmaker*. His proximate target may have been the physicist Fred Hoyle, but his general target was a whole climate of opinion current among mathematicians and physicists.

Von Neumann, in the second place, thought Darwin's theory inadequate. He thought the theory inadequate because the theory did not yet exist. This is as inadequate as it gets. What did exist lacked the fundamentals. It answered no questions. It had no depth. And it was largely anecdotal. This sense of anecdotal has nothing to do with the idea of just-so stories made popular by Lewontin and Gould. Darwin's theory was anecdotal, von Neumann suggested, because it lacked the rich and productive concepts that only mathematicians could provide the sciences.

Writing in the 1967 Wistar Symposium, Murray Eden offered a fine sense of the way in which mathematicians and physicists thought

Darwin's theory inadequate. "The continuity of evolution does not demonstrate that natural laws are operative, for the laws are not known."

Murray then added a most useful analogy. "It is," he wrote referring to Darwin's theory, "as if some pre-Newtonian cosmologist had proposed a theory of planetary motion which supposed that natural force of unknown origin held the planets to their courses."

Just so. This is what Darwin's theory is like; and it was how it appeared to a great many mathematicians and physicists.

And then Murray added a demurral. "The supposition is right enough and the idea of a force between two celestial bodies is a very useful one, but it is hardly a theory."

Far from being controversial among mathematicians in the 1940s and 1950s, a sense of the inadequacy of Darwin's theory was widespread.

And there is a third and final component to von Neumann's skepticism, this one no more than a hint. To say that von Neumann was skeptical of Darwin's theory is not to say that he was a supporter of intelligent design. Yet there is a curious remark he made to Stan Ulam. I suspect that he made the remark at the end of his life. Von Neumann pointed to a house in the distance and remarked to Ulam how absurd it would be to think that the house just assembled itself. The men were discussing Darwin's theory. It was not simply a doubt about improbability to which von Neumann gave voice: it was a more general susurrus of discontent. The remark suggests that just possibly von Neumann's sensibility had undergone a change. Ulam never said anything more to Marco Schützenberger and Marco never said anything more to me.

There remains nonetheless that air of intellectual poignancy. Perhaps von Neumann was aware of his impending death.

So much for what mathematicians thought and think; so much for what von Neumann thought and thank.

I now pass to the point of this exercise. Where did I get my information? Let me tell you. I got my information about von Neumann from the horse's mouth, the horse one step removed from the horse himself.

Quite obviously I did not know von Neumann personally. He was too old and I too young ever to have met. So what I know of views I know at second hand. I know it from my friends.

Stanislaw Ulam was close to von Neumann—very close; and Ulam was also close to Marco and Gian-Carlo Rota—very close again. They were close enough to share their views. I knew Marco and Gian-Co very well; and they were close friends of mine.

Since Marco and I were writing a book together about evolution, our interest in von Neumann's views was natural. Natural, but not consuming. It was a part of the chatter, and it would be wrong to suppose that our curiosity was anything more than curiosity. The subject came up. What had von Neumann thought? We discussed it. Von Neumann stories were told, as they always were.

Let me fix the time and place:

The winter of 1979–1980.

Paris.

And again a year later in Los Angeles at the University of Southern California, where I gave a lecture and Gian-Carlo acted as my host and the source for further stories.

When this issue first emerged, I was asked for references. I am not a von Neumann scholar, and I have not consulted any of the sources. What I know of von Neumann's views, I know from his friends. But what I know is entirely consistent with the development of his own ideas. Von Neumann's work on self-reproducing automata is an interesting attempt to make good the deficiencies of a theory that he understood had deficiencies and needed remedying, a demonstration of the thesis, left unremarked, unanalyzed, and unstated in Darwin's theory, that a self-reproducing automaton is logically possible. It fills one hole.

Ulam's paper "On Some Mathematical Problems Suggested by Problems in Biology" fills another hole. In it he introduced what he called a "biological metric space" into theoretical biology. The date is 1970. The place is the Rockefeller Institute. And the idea, by the way, is pure Marco. Ulam's paper marks his indifference to the dominant paradigm of

Sewell Wright's population genetics. Gone are both Euclidean metrics and Wright's very elaborate apparatus of differential equations.

Whether a theory with quite so many holes in its plush is worth mending is another question entirely.

When the issue of von Neumann's views appeared in the blogosphere, Douglas Theobald, a contributor to Panda's Thumb, much occupied in discharging his indignation, wrote a little post. Von Neumann had apparently said something: "I still somewhat shudder," von Neumann wrote in a letter to George Gamow, "at the thought that highly efficient, purposive organizational elements, like the proteins, should originate in a random process."[1]

Having shuddered, von Neumann apparently got hold of himself. "Yet many efficient (?) and purposive (??) media, e.g., language or the national economy, also look statistically controlled, when viewed from a suitably limited aspect. On balance, I would therefore say that your argument is quite strong."[2]

Theobald suggests that Von Neumann's shuddering remark, taken out of context, "may be the ultimate source of many of the claims that von Neumann was anti-evo."[3]

It may well be, but it is not my source, as I have explained in detail.

Still, what lends to this exchange its lurid aspect is that having shuddered, von Neumann might have kept right on shuddering, for if von Neumann was criticizing Gamow's theories about protein formation, he was right to do so because Gamow's theories were wrong.

And there is the final, the inimitable, touch. I have never claimed that von Neumann was "anti-evo."

What an idea. No one is. What I did claim is that like so many others, von Neumann was profoundly skeptical about Darwin's theory of evolution.

And I have just said why.

# 7. ITERATIONS OF IMMORTALITY

THE CALCULUS AND THE RICH BODY OF MATHEMATICAL ANALYSIS to which it gave rise made modern science possible, but it was the algorithm that made possible the modern world. They are utterly different, these ideas. The calculus serves the imperial vision of mathematical physics. It is a vision in which the real elements of the world are revealed to be its elementary constituents: particles, forces, fields, or even a strange fused combination of space and time. Written in the language of mathematics, a single set of fearfully compressed laws describes their secret nature. The universe that emerges from this description is alien, indifferent to human desires.

The great era of mathematical physics is now over. The three-hundred-year effort to represent the material world in mathematical terms has exhausted itself. The understanding that it was to provide is infinitely closer than it was when Isaac Newton wrote in the late seventeenth century, but it is still infinitely far away.

One man ages as another is born, and if time drives one idea from the field, it does so by welcoming another. The algorithm has come to occupy a central place in our imagination. It is the second great scientific idea of the West. There is no third.

An algorithm is an *effective procedure*—a recipe, a computer program—a way of getting something done in a finite number of discrete steps. Classical mathematics contains algorithms for virtually every elementary operation. Over the course of centuries, the complex (and counterintuitive) operations of addition, multiplication, subtraction, and division have been subordinated to fixed routines. Arithmetic algorithms now exist in mechanical form; what was once an intellectual artifice has become an instrumental artifact.

§

The world the algorithm makes possible is retrograde in its nature to the world of mathematical physics. Its fundamental theoretical objects are symbols, and not muons, gluons, quarks, or space and time fused into a pliant knot. Algorithms are human artifacts. They belong to the world of memory and meaning, desire and design. The idea of an algorithm is as old as the dry humped hills, but it is also cunning, disguising itself in a thousand protean forms. It was only in this century that the concept of an algorithm was coaxed completely into consciousness. The work was undertaken more than sixty years ago by a quartet of brilliant mathematical logicians: Kurt Gödel, Alonzo Church, Emil Post, and A. M. Turing, whose lost eyes seem to roam anxiously over the second half of the twentieth century.

If it is beauty that governs the mathematician's soul, it is truth and certainty that remind him of his duty. At the end of the nineteenth century, mathematicians anxious about the foundations of their subject asked themselves why mathematics was true and whether it was certain, and to their alarm discovered that they could not say and did not know. Caught between mathematical crises and their various correctives, logicians were forced to organize a new world to rival the abstract, cunning, and continuous world of the physical sciences, their work transforming the familiar and intuitive but hopelessly unclear concept of the algorithm into one both formal and precise.

Unlike Andrew Wiles, who spent years searching for a proof of Fermat's last theorem, the logicians did not set out to find the concept that they found. They were simply sensitive enough to see what they spotted. We still do not know why mathematics is true and whether it is certain. But we know what we do not know in an immeasurably richer way than we did. And learning this has been a remarkable achievement, among the greatest and least known of the modern era.

§

Dawn kisses the continents one after the other, and as it does a series of coded communications hustles itself along the surface of the earth, relayed from point to point by fiber-optic cables, or bouncing in a triangle from the earth to synchronous satellites, serene in the cloudless sky, and back to earth again, the great global network of computers moving chunks of data at the speed of light: stock-market indices, currency prices, gold and silver futures, news of cotton crops, rumors of war, strange tales of sexual scandal, images of men in starched white shirts stabbing at keyboards with stubby fingers or looking upward at luminescent monitors, beads of perspiration on their tensed lips. E-mail flashes from server to server, the circle of affection or adultery closing in an electronic braid; there is good news in Lisbon and bad news in Saigon. There is data everywhere and information on every conceivable topic: the way raisins are made in the Sudan, the history of the late Sung dynasty, telephone numbers of dominatrices in Los Angeles, and pictures too. A man may be whipped, scourged, and scoured without ever leaving cyberspace; he may satisfy his curiosity or his appetites, read widely in French literature, decline verbs in Sanskrit, or scan an interlinear translation of the *Iliad*, discovering the Greek for "greave" or "grieve"; he may search out remedies for obscure diseases, make contact with covens in South Carolina, or exchange messages with people in chat groups who believe that Princess Diana was murdered on instructions tendered by the House of Windsor, the dark demented devious old Queen herself sending the order that sealed her fate.

All of this is very interesting and very new—indeed, interesting because new—but however much we may feel that our senses are brimming with the debris of data, the causal nexus that has made the modern world extends in a simple line from the idea of an algorithm, as logicians conceived it in the 1930s, directly to the ever-present always-moving now; and not since the framers of the American Constitution took seriously the idea that all men are created equal has an idea so transformed the material conditions of life, the expectations of the race.

It is the algorithm that rules the world itself, insinuating itself into every device and every discussion or diagnosis, offering advice and making decisions, maintaining its presence in every transaction, carrying out dizzying computations, arming and then aiming cruise missiles, bringing the dinosaurs back to life on film, and, like blind Tiresias, foretelling the extinction of the universe either in a cosmic crunch or in one of those flaccid affairs in which after a long time things just peter out.

§

The algorithm has made the fantastic and artificial world that many of us now inhabit. It also seems to have made much of the natural world, at least that part of it that is alive. The fundamental act of biological creation, the most meaningful of moist mysteries among the great manifold of moist mysteries, is the construction of an organism from a single cell. Look at it backward so that things appear in reverse (I am giving you my own perspective): Viagra discarded, hair returned, skin tightened, that unfortunate marriage zipping backward, teeth uncapped, memories of a radiant young woman running through a field of lilacs, a bicycle with fat tires, skinned knees, Kool-Aid, and New Hampshire afternoons. But where memory fades in a glimpse of the noonday sun seen from a crib in winter, the biological drama only begins, for the rosy fat and cooing creature loitering at the beginning of the journey, whose existence I'm now inferring, the one improbably responding to *kitchy kitchy coo*, has come into the world as the result of a spectacular nine-month adventure, one beginning with a spot no larger than a pinhead and passing by means of repeated but controlled cellular divisions into an organism of rarified and intricately coordinated structures, these held together in systems, the systems in turn animated and controlled by a rich biochemical apparatus, the process of biological creation like no other seen anywhere in the universe, strange but disarmingly familiar, for when the details are stripped away, the revealed miracle seems cognate to miracles of a more familiar kind, as when something is read and understood.

§

Much of the schedule by which this spectacular nine-month construction is orchestrated lies resident in DNA—and "schedule" is the appropriate word, for while the outcome of the drama is a surprise, the offspring proving to resemble his maternal uncle and his great-aunt (red hair, prominent ears), the process itself proceeds inexorably from one state to the next, and processes of this sort, which are combinatorial (cells divide), finite (it comes to an end in the noble and lovely creature answering to my name), and discrete (cells are cells), would seem to be essentially algorithmic in nature, the algorithm now making and marking its advent within the very bowels of life itself.

DNA is a double helix—this everyone now knows, the image as familiar as Marilyn Monroe—two separate strands linked to one another by a succession of steps so that the molecule itself looks like an ordinary ladder seen under water, the strands themselves curved and waving. Information is stored on each strand by means of four bases—A, T, G, and C; these are by nature chemicals, but they function as symbols, the instruments by which a genetic message is conveyed.

A library is in place, one that stores information, and far away, where the organism itself carries on, one sees the purposes to which the information is put, an inaccessible algorithm ostensibly orchestrating the entire affair. Meaning is inscribed in molecules, and so there is something that reads and something that is read; but they are, those strings, richer by far than the richest of novels, for while Tolstoy's *Anna Karenina* can only suggest the woman, her black hair swept into a chignon, the same message carrying the same meaning, when read by the right biochemical agencies, can bring the woman to vibrant and complaining life, reading now restored to its rightful place as a supreme act of creation.

The mechanism is simple, lucid, compelling, extraordinary. In transcription, the molecule faces outward to control the proteins. In replication, it is the internal structure of DNA that conveys secrets, not from one molecule to another but from the past into the future. At some point in the life of a cell, double-stranded DNA is cleaved, so that instead of a single ladder, two separate strands may be found waving gently, like seaweed, the bond between base pairs broken. As in the ancient stories in which human beings originally were hermaphroditic, each strand finds itself longingly incomplete, its bases unsatisfied because unbound. In time, bases attract chemical complements from the ambient broth in which they are floating, so that if a single strand of DNA contains first A and then C, chemical activity prompts a vagrant T to migrate to A, and ditto for G, which moves to C, so that ultimately the single strand acquires its full complementary base pairs. Where there was only one strand of DNA, there are now two. Naked but alive, the molecule carries on the work of humping and slithering its way into the future.

§

A general biological property, intelligence is exhibited in varying degrees by everything that lives, and it is intelligence that immerses living creatures in time, allowing the cat and the cockroach alike to peep into the future and remember the past. The lowly paramecium is intelligent, learning gradually to respond to electrical shocks, this quite without a brain let alone a nervous system. But like so many other psychological properties, intelligence remains elusive without an objective correlative, some public set of circumstances to which one can point with the intention of saying, There, that is what intelligence is or what intelligence is like.

The stony soil between mental and mathematical concepts is not usually thought efflorescent, but in the idea of an algorithm modern mathematics does offer an obliging witness to the very idea of intelligence. Like almost everything in mathematics, algorithms arise from an old wrinkled class of human artifacts, things so familiar in collective memory as to pass unnoticed. By now, the ideas elaborated by Gödel, Church, Turing, and Post have passed entirely into the body of mathematics, where themes and dreams and definitions are all immured, but the essential idea of an algorithm blazes forth from any digital computer, the unfolding of genius having passed inexorably from Gödel's incompleteness theorem to Space Invaders VII rattling on an arcade Atari, a progression suggesting something both melancholy and exuberant about our culture.

The computer is a machine, and so belongs to the class of things in nature that do something; but the computer is also a device dividing itself into aspects, symbols set into software to the left, the hardware needed to read, store, and manipulate the software to the right. This division of labor is unique among man-made artifacts: it suggests the mind immersed within the brain, the soul within the body, the presence anywhere of spirit in matter. An algorithm is thus an ambidextrous artifact, residing at the heart of both artificial and human intelligence. Computer science and the computational theory of mind appeal to precisely the same garden of branching forks to explain what computers do or what men can do or what in the tide of time they have done.

§

Molecular biology has revealed that whatever else it may be, a living creature is also a combinatorial system, its organization controlled by a strange, hidden, and obscure text, one written in a biochemical code. It is an algorithm that lies at the humming heart of life, ferrying information from one set of symbols (the nucleic acids) to another (the proteins).

The complexity of human artifacts, the things that human beings make, finds its explanation in human intelligence. The intelligence responsible for the construction of complex artifacts—watches, computers, military campaigns, federal budgets, this very essay—finds its explanation in biology. Yet however invigorating it is to see the algorithmic pattern appear and reappear, especially on the molecular biological level, it is important to remember, if only because it is so often forgotten, that in very large measure we have no idea how the pattern is amplified. Yet the explanation of complexity that biology affords is largely ceremonial.

At the very heart of molecular biology, a great mystery is vividly in evidence, as those symbolic forms bring an organism into existence, control its morphology and development, and slip a copy of themselves into the future.

The transaction hides a process never seen among purely physical objects, one that is characteristic of the world where computers hum and human beings attend to one another. In that world intelligence is always relative to intelligence itself, systems of symbols gaining their point from having their point gained. This is not a paradox. It is simply the way things are. Two hundred years ago the French biologist Charles Bonnet asked for an account of the "mechanics which will preside over the formation of a brain, a heart, a lung, and so many other organs." No account in terms of mechanics is yet available. Information passes from the genome to the organism. Something is given and something read; something ordered and something done. But just who is doing the reading and who is executing the orders, this remains unclear.

# 8. A Long Look Back: The Wistar Symposium

THE WISTAR SYMPOSIUM WAS HELD IN APRIL OF 1966. THE MEETING reflected a certain discontent that had been simmering in biology for many years.[1] In his opening remarks, Peter Medawar put the facts of the matter as plainly as possible: "The immediate cause of this conference is a pretty widespread sense of dissatisfaction about what has come to be thought as the accepted evolutionary theory in the English-speaking world, the so-called neo-Darwinian theory. These objections to current neo-Darwinian theory are very widely held among biologists generally; and we must on no account, I think, make light of them."

It is worth observing, with a gathering sense of embarrassment, what a difference the intervening years have made in the life of our own scientific culture. No prominent biologist would today dream of expressing a sense of dissatisfaction about evolutionary theory unless, of course, it were in the context of remarks calling for an increase in its level of funding. The symposium did not lack for notable figures in mathematics, biology, physics, and computer science: Murray Eden, M. P. (Marco) Schützenberger, Stanislaw Ulam, Richard Lewontin, Sewall Wright, Peter Medawar, C. H. Waddington; but it was Eden and Schützenberger who most consistently expressed the widespread sense of dissatisfaction to which Medawar had alluded.

These were distinguished men. Eden held a professorship in electrical engineering at MIT, and Schützenberger, with a degree in medicine as well as mathematics, had been a visiting member of the faculty at the

Harvard Medical School; he was a permanent professor of mathematics at the University of Paris (Jussieu). Among mathematicians, he was well known for his work on the algebraic theory of words, and among mathematical linguists, for the Chomsky-Schützenberger representation and enumeration theorems. The specific arguments that Eden and Schützenberger advanced in Philadelphia were of their time and place. How could it be otherwise? Arguments are one thing; the problems to which they are addressed, quite another. Both Eden and Schützenberger had for a time a kind of second sight: they could see problems that other participants could not see at all. These problems remain today everywhere in evidence, nowhere in sight.[2]

§

Finding a needle among hayseeds is no difficult task. One of the hayseeds is bound to prick himself. Finding a needle in a haystack is proverbially difficult. The larger the haystack, or the smaller the needle, the harder the ensuing problem. The needle is, *entre nous soit dit*, something of an irrelevance. The problem is one of finding a particular stalk in the stack. If all stalks are identical, there is no problem. It is when one stalk is bent, and the others not, that a problem emerges.

Haystacks are put together by heaping up hay stalks; English sentences, by concatenating words; and English words, by concatenating letters. A few definitions now follow. An alphabet, A, consists of a finite set of discrete, letter-like objects. English, Greek, and Arabic have alphabets all their own, but the nucleic and amino acids, Murray and Marco both argued, are also alphabetic. The words W over A are formed by a binary associative operation, *Spot* emerging from A = {S, P, O, T} as a series of stutters: **S, SP, SPO, SPOT**. A finite and discrete alphabetic system, $S_A$, is a pair <A, W>, letters on the left, words on the right. There are $n!$ permutations among the $n$-letter words; but the combinations are exponential in $k^n$, where $k$ designates the number of elements in A. Its reputation for comparative pokiness notwithstanding, the factorial function grows faster than the exponential function. For sufficiently large $n$,

$$\frac{a^n}{n!} < \epsilon.$$

With no specification of $n$ or $k$ or even the relationship between them, $S_A$ remains an abstraction, like the Form of the Good.

The discrete alphabetic systems are discrete all the way down. It is always possible to dissect a word into a finite number of discrete letters. Not every alphabetic system is discrete. Chinese calligraphy offers a case to the contrary. The very beautiful character

may be dissected into a finite number of primitive strokes, but strokes are sets of similar *shapes*; and for the analysis of similarity in shape, something like a tolerance space is needed.

Suppose that A = {*Spot, see, run*}, and W is the set of all distinct three-letter words on A. *Distinct*, note. Permutations are in power. *See Spot run* is in, but *see see see* is out. Nothing could be simpler, and nothing is. Within the ambit of English speakers, *See Spot run* is an invitation, or an injunction, to look at Spot run. The word captures what the picture in primers illustrates. The word makes perfect sense,[3] one reason, I suppose, that children are encouraged to see Spot run. One among six permutations has a property that the others lack—a *discernible* feature. It is a curiosity of such systems that their features may be discerned before they are, if ever, defined.

Douglas Axe has remarked of artifacts that they can be spotted a mile away. No derivation is needed. These are views that may be traced back to long-forgotten British intuitionists. "The sense of obligation to do, or of the rightness of, an action of a particular kind," H. A. Prichard remarked, "is absolutely underivative or immediate."[4] This is one way of putting things. René Thom offered another. Discernible and isolated properties, he called *saliences*. They arise up from some underground Euclidean space: "Les pré-gnances sont des actions propagatives émises par les formes saillantes, qui investissent ces formes, et cet investissement provoque dans l'état de ces formes des transformations appelées effets figuratifs."[5]

I once thought to ask Thom what this meant. "If there were nothing to notice in Nature," he answered, "no one would notice anything."

If it is hard to fault this observation, it is equally hard to say *why* it is true. All electrons are identical: they are equally manifestations of the same quantum field. The natural numbers are distinct in respects that go far beyond size. Any two consecutive natural numbers are by defini-tion on opposing sides of the mathematical aisle; and four or more natu-ral numbers are known to give rise to political factions. The individuality of the natural numbers is the more remarkable in virtue of the fact that every natural number beyond 1 is either a prime number or can be writ-ten as a product of positive prime numbers:

$$n = p_1^{a_1} p_2^{a_2} \dots p_k^{a_k} = \prod_{i=1}^{k} p_i^{a_k}.$$

This compelling variety—what is its source?

§

One and the same system may possess a discernible feature in one con-text, but not in another. One sees Spot run in English, but not in French. French readers see only so many hunched-over letters. Those in the know see Spot run. Chemistry treats the enormous variety of chemical compounds as words drawn from a finite alphabet for which the peri-odic table of the elements provides the letters. Organic chemists see in chemical words a scheme in which Carbon is above all the other ele-ments. Quantum physicists regard the periodic table of the elements as an adaptive mask, one that when withdrawn reveals the alien apparatus of quantum field theory. Organic chemistry is distinctive in their eyes only to the extent that the quantum calculations it demands are impossible to enforce. The hydrogen atom, H, with its strabismic electron, represents the great triumph of quantum mechanical methods; but $H_2O$ remains a tormenting hair shirt of a compound.[6]

§

Discrete mathematics stands to continuous mathematics in roughly the proportions of one to ten. Impressed by the difference, a number of mathematicians have asked whether they might not get rid of the alphabetic systems altogether. It is for this reason that Steve Smale, Martin Shub, and Lenore Blum have argued that the digital computer should be treated as a continuous input-output map defined over the real numbers. Real numbers they accept as ineliminable idealizations, like point-masses in Newtonian mechanics. The familiar Turing machine, having been dissolved and then reimagined as a butterfly's pharate, emerges after pupation as the Blum-Shub-Smale machine. Nothing is lost. The classical architecture of twentieth century logic stands undefiled. Theorems remain theorems. Many alphabetic concepts could be reinterpreted in this way. Those saliences to which Thom appealed might be given continuous standing as singularities in the space of smooth maps. This is precisely how Thom *did* interpret them, although his classification theorem made use of germs rather than functions.[7] Sard's theorem suggests just why the singularities of a smooth map should be isolated. The set of singular values of a smooth function $f$ from one Euclidean manifold to another is of measure 0. This is a generic property. The singularities are discernible because they are singular. No one can miss the change in $x^3$ at the origin. They are isolated because the set of them is small.

When I mentioned Smale's work to Marco, he remarked that it was insane. His view of Thom's theory was no more forthcoming.

§

The specific system in which Spot was recently seen running departs from the general run of alphabetic systems in that only one among its six word orders makes sense. There is no additional concourse between words. Properties such as making sense are imposed on words from the outside, like the property of being the winning hand in a game of poker. This is true of the representation groups in mathematical physics as well. In particle physics, the group $SU(2)$ is commonly mapped to the up and down quarks. Group representation theory imposes on $SU(2)$ a role in the description of quarks, but the imposition is not intrinsic. In a world without quarks, $SU(2)$ would be just another group, a lurker.

A given alphabetic system is what it is: What are *its* properties? That is one question. But systems belong to families. What are *their* properties? That is another, an intramural, question.

The discernible systems are themselves discernible. This is a point that Marco made in Philadelphia, and one that went unremarked. Most finite alphabetic systems enjoy an uninflected existence. There is nothing special about them. What is so very curious about a needle in a haystack is the fact there exist haystacks that contain needles. So many haystacks do not. The movement in thought is now opposite to that taking place in theoretical physics, where any number of specific cases are subordinated to the generic concept of a symmetry group. A physician, as well as a mathematician, Marco understood that living systems are quirky, isolated, idiosyncratic, strange, bizarre, unusual, peculiar, and odd. Anyone familiar with the astonishing cases that Michael Denton has collected in *Evolution: Still a Theory in Crisis*, will know what Marco meant.[8] The finite alphabetic systems are strongly discernible in the great sea of purely physical systems.

There is nothing else like them.

§

The finite alphabetic systems stand in the shadow of systems that are infinite and discrete. These are structures that for the most part have a representation in the natural numbers—the pair $\mathbb{N} = \langle \emptyset, S \rangle$, where $\emptyset$ is the empty set, and $S(a) = a \cup \{a\}$ for every set $a$. Marco (but not Murray) believed that the division between the real and the natural numbers was absolute, no matter the fact that the real numbers may ultimately be defined by such devices as Dedekind cuts or partitions. The real numbers are uncountable and the natural numbers, countable.

The physical properties of the universe, Marco argued, are describable in terms of the real and complex numbers; *everything else*, in terms of the natural numbers. The rational numbers Marco regarded as a gift.

The prime number theorem offers an instructive example of an infinite discrete system with strongly discernible properties. There are four prime numbers less than ten. They are 2, 3, 5, and 7. There are eight prime numbers less than twenty; twenty-five prime numbers less than one hundred; and 168 prime numbers less than 1,000. A cursory look at the ratios 4/10, 8/20, 25/100, and 168/1,000 might suggest that as N is getting larger, the number of primes relative to N is getting smaller. This is the burden of the prime number theorem. Let $\pi(N)$ be the prime counting function. For any natural number, $\pi(x)$ returns the number of primes that are less than $x$. The number of primes less than $x$ is roughly equal to the ratio of $x$ to its natural logarithm: $\pi(x) \sim x/\log(x)$. For a large enough value of $x$, the probability that an integer chosen at random is a prime number is very close to $1/\log(x)$:

$$\lim_{x \to \infty} \frac{\pi(x)}{x / \log(x)} = 1$$

where $\pi(x)$ is the prime counting function.

This is a theorem in which every possible parameter is precisely controlled. The definition of a prime number is both explicit and constructive. For any given $n$, trial by division, although slow, sooner or later determines whether $n$ is prime. The prime counting function $\pi(x)$ describes what is being counted. The fact that $x \to \infty$ shows where $x$ is going.

It does not in the least matter how fast $x$ is approaching a limit. The conclusion is the same. At a certain point, it becomes fruitless to search for prime numbers by chance.

At *what* point? This the prime number theorem does not say. The correct answer is also the only answer: it all depends.

The prime number theorem, Marco often observed, conveys a message, one that is morally true, as physicists sometimes say. Some things tend to get lost as their underlying space gets larger and larger. Their isolation becomes more and more pronounced. This is neither a law of mathematics nor a law of nature. It is not true for every property. The even numbers do not begin to diminish themselves as $n$ gets larger and larger. But it *is* true with respect to the prime numbers. Up to any finite $n$, the composite numbers, $\gamma(x)$, are thicker on the ground than the prime numbers $\pi(x)$, and if in some game, the competition hinges on whether $\pi(x) > \gamma(x)$, or the reverse, the game would generically be decided in favor of the composite numbers.

If the composite numbers are accepted as a mathematical metaphor for failure, the prime number theorem establishes the old human truth that it is easier to fail than to succeed. It does not say *why*.

§

The finite alphabetic systems are what they are: They are finite. They can be arranged by the increasing order of their alphabets: $A_1$, $A_2$, ... , $A_n$ ... Alphabets under this scheme are growing one by one, or, at most, by a linear factor $n = n - 1 + b$. The words W drawn on a finite alphabet are otherwise. They multiply like mad. Combinatorial concepts are in charge. There are $n!$ permutations among the $n$-letter words; and $k^n$ combinations. Alphabets are growing slowly; words are growing explosively, so much so that the ratio $A_n/W_n$ tends to zero. Such is *combinatorial inflation*. Something is getting bigger fast, but unlike blow-ups in cosmology, combinatorial inflation is a matter entirely of permutations and combinations, the way in which things are arranged.

Combinatorial inflation comes to exert its influence as a concept when the discernible *words* are recalled from obscurity. How do sensible words like *see Spot run* grow in the general growth spurt among words? Like A? That is one possibility. Or like W? That is another possibility. At the Wistar Symposium, both Murray and Marco affirmed that the discernible words grow like A and *not* like W. The meaningful words grow in proportion to their alphabet, and not in proportion to words in their family environment; so do the functional proteins; so do artifacts of almost every variety; so, too, most things of beauty; and complicated machines, especially biological machines such as the bacterial flagellum.

Down at the bottom of the evolutionary chain, where life emerged from the primordial goo, the distinction between those systems that were alive and those that were simply hanging around to no good purpose may have been very elusive. Perhaps a rough parity between the living and the dead prevailed. Up at the top of the evolutionary chain, nothing like this is remotely true. Even if once small, the ratio between what is living and what is dead is now large. There are many more ways in which to realize a composite number than a prime number; and many more ways in which to realize the dead than the living.

It is fair to say that we knew this all along, but if this is so, both Murray and Marco argued, it is hard to say why Darwinian thought should have enjoyed its longstanding success.

§

If failures are in general polygamous and success monogamous, how were the discernible elements in any system ever found? To say that Nature has found them is true but irrelevant. Strongly discernible properties may be recognized when they are found, but this says nothing about finding them before they are recognized. Whatever the specific details, combinatorial inflation encourages an invocation of Borel's law of large numbers. If an event $E$ occurs $X_n$ times in $n$ trials, then the probability $p$ of $E$ is

$$p = \lim_{n \to \infty} \frac{X_n(E)}{n}.$$

It is worthwhile, if only as an aside, to reject the surprisingly common misconception that improbable events happen *all* the time. In time snatched from *geschäft*, physicists often make very similar claims, especially when they are assuring one another that creationism is a form of folly. If unlikely events happened all the time, they would not be unlikely. On the contrary. Unlikely events happen as often as the laws of probability suggest. The thesis that for all times, there is *some* unlikely event that is apt to occur is trivial; and the thesis that for some unlikely event, *it* happens all the time is silly.

The participants at the Wistar Symposium, the brilliant mathematician Stanislaw Ulam among them, found it is astonishingly easy to miss the significance of these remarks. Combinatorial inflation, together with Borel's law of large numbers, leads to two surprising inferences.

The first: that under certain circumstances, an appeal to chance is pointless.

The second: that those circumstances are surprisingly common.

§

In his talk at the Wistar Symposium, Murray managed to create an attractive elementary argument about protein evolution. His argument owed *nothing* to the rich and sophisticated systems of population genetics, but like the living dead, Murray's argument has made its disturbing appearance in the literature well after the Wistar Symposium had been brought to a close. Writing in 1989, the population geneticist Daniel Hartl offered a retrospective marked most obviously by bafflement that the thing was still there:

> The evolution of novel catalytic activities was well recognized as paradoxical. As expressed by Eden… in the Wistar Symposium on *Mathematical Challenges to the Neo-Darwinian Interpretation of Evolution*, the problem was in the infinitesimal probability that catalytically useful proteins containing hundreds of amino acids could result from the random assembly of their amino acid subunits. For example, the probability that a particular sequence of 100 amino acids in a functional polypeptide would occur by chance combination is only $20^{-100}$… which, even allowing for some freedom in the amino acids that can occupy individual

positions, is pretty small. *While the probability paradox was not emphasized in evolutionary thinking, it remained unresolved in the major theories of enzyme evolution,* including the classical theory of duplication and divergence in which new catalytic activities were supposed to evolve by random amino acid changes resulting from nucleotide substitutions in duplicate copies of preexisting genes.[9]

I added italics to the first part of the final sentence of this quotation to stress its origins: it has come straight from the horse's mouth.

That horse having spoken, he required only a few more words to reach an impediment in evolutionary thought. It was precisely the impediment to which Murray called attention at the Wistar Symposium. "Either functionally useful proteins are very common," Murray argued, "… so that almost any polypeptide" of random amino acid sequence "has a useful function to perform, or else... there exist certain strong regularities for finding useful paths" in protein evolution.

Alone among the participants at the Wistar Symposium, Sewall Wright remained unvexed by this argument. Comparing the discovery of a particular protein to a game of Twenty Questions, he observed that five hundred *yes* or *no* questions could certainly serve to specify a particular protein one hundred amino acids in length. In this, he was correct. For every possible open position in a chain of amino acids, five questions, at most, are required to identify any particular amino acid. The questions proceed by successively reducing twenty possibilities to a single choice by partitioning twenty in five ways.

If Nature had the wherewithal to interrogate a partition five ways, why, one wonders, might she not have skipped the whole business and simply specified precisely the right one hundred amino acids in the first place? The problem with Wright's suggestion is obvious. In order to ask the question, someone must know the answer. Neither Wright nor anyone else has ever suggested who this might be.

Daniel Hartl was entirely correct. The paradox to which Murray alluded *is* unresolved. Chance is of no relevance whatsoever; and Darwin's theory is entirely a stochastic theory, one in which fitness, $f(X_1, X_2)$, is determined by the product of two random variables, the first denoting random variations in the genome, the second, random variations in the environment. The product of two random variables is again a random variable.

If not chance, then what? If functional proteins are isolated in the space of all possible proteins, the topology of the space must play a role in their discovery.

Both Murray and Marco argued that this must be true.

§

A topological space, S, comprises a pair $\{X, \Omega\}$, where X is a set, and $\Omega$ a collection of subsets on X. There are three axioms: i) both the empty set and X itself belong to $\Omega$; ii) any finite or infinite union of sets in $\Omega$ belong to $\Omega$; and iii) the finite intersection of sets $\Omega$ is in $\Omega$. The elements of $\Omega$ are open sets; $\Omega$ is a topology on X.

Certain topologies, $\Omega$, topologists say, are path-wise connected. The name reveals its own identity. In such topologies, there are paths between points. It is possible to get around. Let U be an open set and $f$ a continuous function from [0, 1] to U. The topology, $\Omega$, is path-wise connected if for any two points $r$, $q$ in U, there is a path $f(x)$ in U between $r$ and $q$ such that $f(0) = r$ and $f(1) = q$.

The distinguished point topology enlarges on this idea. Let X once more be a set, and $p$ a particular point in X. The collection of subsets S in X such that either $p$ belongs to S or S is empty comprises a topology on X. The open sets are the subsets S. X may be finite, countably infinite, or uncountably infinite.

Two remarkably interesting consequences follow from these definitions. The first is entirely obvious. The complement of the open sets could not contain any interior points. An interior point of a given subset of X must be embedded in an open set. And the open sets by definition contain $p$.

The second is less obvious. Distinguished point topologies are path-wise connected. Since U is open, it contains $p$. So for any point $r$ in U, there is a path from $p$ to $r$.

Topologies that are path-wise connected, and particular point topologies, play a strange role in discussions of protein evolution. The metaphor by which the nucleic acids and the proteins were imagined as finite and discrete alphabetic spaces has always suffered an internal weakness. It is this. If proteins evolve in the fashion of the metaphoric chain

WORD $\rightarrow$ WORE $\rightarrow$ GORE $\rightarrow$ GONE $\rightarrow$ GENE,[10]

protein chemists would be well-satisfied and the metaphor would have appreciated in value from the obvious relevance of proximate examples.

What of the sequence that begins with SYZYGY?

A path-wise connected topology, Murray might have argued, is no luxury; if it is needed to explain the facts of protein evolution, it is also needed to redeem a metaphor that has dominated molecular biology since, at least, the discovery of DNA by Watson and Crick.

But if path-wise connected topologies are required in biology, who ordered that?

§

The whale is unusual among the fish in that, first, it is not a fish, and second, it has in Pakicetus an imagined ancestor that lived entirely on land. If the transition from Pakicetus to the modern whale has always seemed as streamlined as the whale itself, this is, in part, because evolutionary biologists have attended to the streamlining at the expense of the blubber. (On a very common view among paleontologists, Pakicetus, the land-dwelling ancestor of the whale, became the modern whale in something like eight million years, but recent research seems to suggest that, having for entirely inexplicable reasons decided to leave the land, Pakicetus had only a million years to become entirely aquatic.[11]) Evolutionary sequences such as this, Marco argued, are both stable and oriented; and although very suggestive, his argument was, and has remained, almost entirely anecdotal. They are stable because in their trajectories, they obviously remained indifferent to small perturbations; and oriented, because it is easy to fix an origin, an end point, and a direction between the two. Looking at the splendid illustrations of whale evolution in which any number of whale-like creatures are stacked from the bottom to the top, it is all too easy to assign to the bottom-dwellers an intense desire to get to the top.

Every living creature, Marco argued, is the instantiation of two alphabetic systems. The first is made up of the nucleic acids, or the proteins; the second, of the organism itself, the full three-dimensional creature—the lion, the spider, the human being. To each system, Marco proposed to assign its own natural metric topology. In 1966, the genome was an object relatively new in anyone's experience; and both Murray and Marco assumed that since it appeared to embody a finite and discrete alphabetic system, this is what it must be. This metaphor survives throughout the biological science.

Suppose that a genome, G, embodies a discrete and finite alphabetic space; and suppose that it has, in addition, a metric structure. And why not? The metric spaces are among the topological spaces. They simply have a more determinate quantitative character. There thus exists a distance function $d_G(w_1, w_2)$ defined on the words, **W**, of G. It is $d_G(w_1, w_2)$ that measures the number of changes needed to bring $w_1$ into alignment with $w_2$. The Hamming metric is the most natural way to define $d_G(w_1, w_2)$.

Whatever else G may be doing, it is changing in time; and it therefore has the additional structure of a dynamical system. Changes in G are governed by a probability transition system, **PTS**. Given a word $w_i$, Pr $(w_1, w_k)$ designates the probability that $w_i$ will change to $w_k$. A **PTS** with respect to G and its natural metric work hand in glove. It is more likely that a word will not change at all than that it will change with respect to one letter; more likely that it will change with respect to one letter than with respect to two, and more likely still that it will change with respect to $n - i$ letters than to $n$ letters. The further the distance between words, the less the likelihood of a transition between them. For any three words $w_i$, $w_k$, and $w_j$,

$$d_G(w_i, w_k) < d_G(w_i, w_j) \Leftrightarrow \mathrm{pr}(w_i, w_j) < \mathrm{pr}(w_i, w_k).$$

This principle is true of English words as well. It is more likely that the word WORE changes to the word GORE than that it changes to the word DOOR. The first two words are closer to one another than either of them is to the third.

One space having been given, another is about to be gotten. It is the space, $P = \{p_1, p_2, \dots, p_n\}$ of phenotypes corresponding to the genotypes, G, under some complex mapping $f$. P is a finite dimensional Euclidean vector space with its usual norm and product. This is the more or less standard structure assigned by population geneticists to population genetics. There is thus a distance function $d_P(p_1, p_2,)$ defined on P.

If there is a metric on P, and another on G, there exists, as well, an induced metric, $d_{GI}$, on G. For any two elements $x$ and $x'$ of P, the distance $d_{GI}(w, w')$ between the words $w$, $w'$ is defined by the inverse of $f$ in G.

$$d_G\left(f^{-1}p, f^{-1}p'\right) = d_{GI}(w, w').$$

A **PTS** is engaged on the level of G, and its actions are followed as a stable and oriented sequence in P. One whale after the other is making for the open ocean. The question that Marco asked at the Wistar Symposium is just which metric is in charge of this sequence? It cannot be $d_G(w_1, w_2)$ because under the Hamming metric, an arbitrary **PTS** could not possibly realize a stable and oriented system.

It must, therefore, be the induced metric.

But this is simply a variant of Murray's argument, one gaining life as a series of assumptions in metric topology rather than any distinguished point topology.

No matter the topological details, how were the relevant topologies found; and if not found at all, how imposed?

And by whom?

This argument and the question it provokes have enjoyed a second life in the literature of mathematical biology. Thus Michael Gromov, the distinguished mathematician and Fields medalist, writing with Alessandra Carbone, comments:

> The subject matter of taxonomy is constructing various metrics on the spaces of genotypes, species or other biological and biochemical entities (such as proteins and RNAs). There are two different, mathematically dual, approaches introducing a metric in the (moduli) space of structured objects. The first is the *evolutionary* (cladistic) approach, where the (phyletic) metric is defined by the length of the shortest path of elementary modifications of the object, similarly to the construction of intrinsic metrics by Gauss-Riemann-Kobayashi. Limiting such a metric to the space of actually existing *biological* entities often exhibits a pronounced tree-like behavior. This is due to the huge size of the space of possible structures, e.g. of bp sequences of length $10^9$, where the branches of the evolution process are very unlikely to come together and form cycles. The second (phenetic) metric is the *phenomenological* one, similar in spirit to the Caratheodory metric on complex spaces. Here, we pick up some distinguished physiological parameters, e.g. size, style of breathing and nutrition, reproduction patterns, etc., and thus map our space into the space of parameters, where we choose some simple metric and induce it back to our original space. The dream of biologists is to independently construct these two metrics such that they would become equal on the space of the existing organisms.[12]

§

Certain concepts, and with them, certain arguments, are destined to the double life. They appear first in the professional literature, and they appear again in the popular imagination. The result is sometimes an embarrassment, as when some thoroughgoing primitive is pleased to assign his work to Gödel's incompleteness theorems, but this is not always the case. Turing computability is a hard, precisely defined mathematical concept; but the concept of computation itself is now a part of a culture larger than mathematics itself. It has not been widely misused; it is not like quantum entanglement. Combinatorial inflation is among the happy concepts. Its technical aspects are hardly trendy, and the essential combinatorial theorems, lemmas, and definitions have been a part of mathematics since the eighteenth century. What is new is the noticing of it; and once it has been noticed, as both Murray and Marco did in Philadelphia in the spring of 1966, the noticing itself leads naturally to an immensely pertinent argument—the argument from combinatorial inflation.

It is an argument best cast as a destructive dilemma. Consider a discernible property $\zeta$ of the words in some finite and discrete alphabetical system S. Now either $\zeta$ is rather like A, its alphabet, or like W, its words. If it is like W, $\zeta$ should be common among the words in S, something as easily found as noted; and if it is like A, isolated among the words in S, as difficult to find as it is easy to notice.

If $\zeta$ is common, this invites an empirical question: Where is it?

And if it is not common, it invites a philosophical question: Why not?

Writing in *Nautilus*, astronomer Caleb Scharf has argued that "if machines continue to grow exponentially in speed and sophistication, they will one day be able to decode the staggering complexity of the living world, from its atoms and molecules all the way up to entire planetary biomes."[13]

Computers *are* getting faster. According to Moore's law computer processing speed doubles every two years: $P = P_0 \times 2^n$. But speed is one thing, sophistication quite another. Whatever Scharf may imagine computers doing in the far future, in the here and now their sophistication has grown by roughly a linear factor. Improvements are real, but there is nothing exponential about them.

An application of the dilemma now follows. If computer sophistication proceeded as rapidly as computer speed, we would expect to see very sophisticated systems everywhere. This is obviously not the case.

Why not? There is no need for any very elaborate argument. We do not see computers nearly as sophisticated in most cognitive tasks as a three-year-old child, or an otherwise intelligent ape, for that matter, because we have not the requisite scientific theory to explain what they do.

The Copernican Principle amounts to the assertion that because there are no distinguished coordinates anywhere in the physical universe, Earth is not in any relevant sense special, marked, or distinguished. The principle has afforded physicists the satisfaction of self-abnegation without any of its attendant discomforts. Copernicus is often held to be the first in a long line of Western thinkers whose effect has been chiefly to depreciate Western thinkers. It is a category in which Darwin is supreme.

A Bad Thing is about to happen to the Copernican Principle. In 1961, Frank Drake published an equation that would ostensibly serve to indicate the number of advanced civilizations astronomers might expect to find in the Milky Way galaxy. The equation was a masterpiece of mathematical mummery:

$$N = R \times f_p \times n_e \times f_l \times f_i \times f_c \times L,$$

where R is the rate of star formation, $f_p$, the fraction of stars that form planets, $n_e$, the number of Earth-like planets, $f_l$, the fraction of Earth-like planets on which life emerges, $f_i$, the fraction of planets on which intelligent life emerges, $f_c$, the fraction of planets capable of achieving interstellar communication, and L, finally, the length of time such a civilization might remain detectable.

Well before Drake published his equation, Enrico Fermi, contemplating the possibility of life beyond the Earth, asked a simple, pointed, and devastating question: *So where are they?*

No one knows; no one has ever detected the slightest suggestion of life beyond Earth. The Bad Thing now follows in the form of a dilemma. Either Earth is unique, after all, or life must be quite common in the cosmos. It is not common at all. Earth must, therefore, be unique.

If this is so, *why* is it so?

This argument and this question have, I suggest, a very large relevance; it has become mysteriously pertinent. That this is so represents the real achievement of the Wistar Symposium.

# III. Microbe to Man

# 9. Responding To Stephen Fletcher's Views in the *Times Literary Supplement* on the RNA World

To the Editor

*The Times Literary Supplement*

Sir—

Having with indignation rejected the assumption that the creation of life required an intelligent design, Mr. Fletcher has persuaded himself that it has proceeded instead by means of various chemical scenarios.

These scenarios all require intelligent intervention. In his animadversions, Mr. Fletcher suggests nothing so much as a man disposed to denounce alcohol while sipping sherry.

The RNA world to which Mr. Fletcher has pledged his allegiance was introduced by Carl Woese, Leslie Orgel, and Francis Crick in 1967. Mystified by the appearance in the contemporary cell of a chicken in the form of the nucleic acids, and an egg in the form of the proteins, Woese, Orgel, and Crick argued that at some time in the past, the chicken was the egg.

This triumph of poultry management received support in 1981, when both Thomas Cech and Sidney Altman discovered the first of the ribonucleic enzymes. Their discoveries moved Walter Gilbert to declare the existence of an RNA world in 1986. When Harry Noeller discovered that protein synthesis within the contemporary ribosome is catalyzed by ribosomal RNA, the existence of an ancient RNA world appeared "almost certain" to Leslie Orgel.

And to Mr. Fletcher, I imagine.

If experiments conducted in the here and now are to shed light on the there and then, they must meet two conditions: they must demonstrate in the first place the existence of a detailed chemical pathway between RNA precursors and a form of self-replicating RNA; and they must provide in the second place a demonstration that the spontaneous appearance of this pathway is plausible under prebiotic conditions.

The constituents of RNA are its nitrogenous bases, sugar, and phosphate. Until quite recently, no completely satisfactory synthesis of the pyrimidine nucleotides has been available. The existence of a synthetic pathway has now been established. (Matthew W. Powner, Béatrice Gerland, and John D. Sutherland, "Synthesis of Activated Pyrimidine Ribonucleotides in Prebiotically Plausible Conditions," *Nature* 459 (2009), 239–242). Questions of prebiotic plausibility remain. Can the results of Powner et al. be reproduced without Powner et al.?

It is a question that Powner raises himself: "My ultimate goal," he has remarked, "is to get a living system (RNA) emerging from a one-pot experiment."

Let us by all means have that pot, and then we shall see further.

If the steps leading to the appearance of the pyrimidines in a prebiotic environment are not yet plausible, then neither is the appearance of a self-replicating form of RNA. Experiments conducted by Tracey Lincoln and Gerald Joyce at the Scripps Institute have demonstrated the existence of self-replicating RNA by a process of in vitro evolution. They began with what they needed and purified what they got until they got what they wanted.

Although an invigorating piece of chemistry, what is missing from their demonstration is what is missing from Powner's and that is any clear indication of prebiotic plausibility.

I should not wish to leave this discussion without extending the hand of friendship to every party.

Mr. Nagel is correct in remarking that Mr. Fletcher is insufferable. Mr. Walton is correct in observing that the RNA world is imaginary. And Mr. Fletcher is correct in finding the hypothesis of intelligent design unacceptable.

He should give it up himself and see what happens.

David Berlinski
Paris

# 10. The Activity of a Cell Is Like That of a Factory

IT IS ALWAYS INTERESTING TO SEE HOW OBSERVATIONS THAT WE HAVE been making for years are now a part of the general chatter. From "Science without Validation in a World without Meaning," by Edward R. Dougherty, Distinguished Professor of Electrical Engineering at Texas A&M, writing in the journal *American Affairs*:

> At the cellular level, biology concerns the operation of the cell in its pursuit of life, not simply the molecular infrastructure that forms the physiochemical underpinnings of life. *The activity of a cell is like that of a factory, where machines manufacture products, energy is consumed, information is stored, information is processed, decisions are made, and signals are sent to maintain proper factory organization and operation.* Once a factory exceeds a very small number of interconnected components, coordinating its operations goes beyond a commonsense, nonmathematical approach. Cells have massive numbers of interconnected components.[1] [Emphasis added.]

No one bats an eye when something like this is said in *American Affairs*; but let it be said on *Evolution News* and the suspicion at once bursts into flower that the devil himself is at work, his forked tail frisking.

# 11. A GRADUATE STUDENT WRITES

Editor's Note: Shortly after Stephen Meyer's *Darwin's Doubt* was published in June 2013, *Panda's Thumb*, a blog dedicated to the defense of evolutionary theory, published a lengthy critique by Nick Matzke. Part of Berlinski's response, below, followed in July of that year.

NICK MATZKE HAS WRITTEN A CRITIQUE OF STEPHEN MEYER'S *Darwin's Doubt*. Having for years defended Darwin's theory as an employee of the National Center for Scientific Education, Matzke has determined to learn something about the theory as a graduate student at the University of California, an undertaking in the right spirit but the wrong order. Would that he had done things the other way around. His animadversions are written with all of the ebullience of a man sure enough of his conclusions not to worry overmuch about his arguments. They are wrong in the small, wrong in the large, and wrong all around. A pity. The Darwinian establishment is hardly without resources of its own, and had Matzke devoted more thought to his critique, he might have spared us the embarrassment of improving his arguments before rejecting his conclusions.

*Darwin's Doubt* advances three theses: First, that the Cambrian explosion is a real event; second, that the Cambrian explosion has not been explained by neo-Darwinian or other evolutionary mechanisms; and, third, that the Cambrian explosion is best explained by an inference to intelligent design. Matzke, in his critique of Meyer, concentrates on the first of these theses, barely mentions the second, and fails to engage Meyer's arguments for the third.

*Darwin's Doubt* makes its case for the reality of the Cambrian explosion chiefly, but not entirely, on the basis of the fossil record. Paleontology has pride of place. It is where the bodies are. Representatives of twenty-three of the roughly twenty-seven fossilized animal phyla, and the roughly thirty-six animal phyla overall, are present in the Cambrian fossil record. Twenty of these twenty-three major groups make their appearance with no discernable ancestral forms in either earlier Cambrian or Precambrian strata. Representatives of the remaining three or so animal phyla originate in the late Precambrian, but they do so as abruptly as the animals that appeared first in the Cambrian, and they lack clear affinities with the representatives of the twenty or so phyla that first appear in the Cambrian.

An account of their appearance must logically be focused on either the earliest part of the Cambrian or the Precambrian. Where else to look? But in looking there, Meyer argues, there is nothing much to see. Not nothing, of course. The well-known sequence that begins with the acritarchs and gutters into the small shelly fauna is an example, one in which Matzke invests his hopes without hedging his bets. "The earliest identifiable representatives of Cambrian 'phyla,'" Matzke writes, the twitch of misplaced quotation marks around a word that does not need them, "don't occur until millions of years after the small shelly fauna have been diversifying."[1] But while the small shelly fauna offer something to see, they reveal nothing of interest. No paleontologist believes that the small shelly fauna are ancestral to all the Cambrian phyla.

Trilobites are an example. These strange and complex creatures, their eyes staring hypnotically, appear during the early Cambrian, in the Atdabanian stage. Having quite obviously gotten to where they appear in the fossil record, how did they get there? One speculative scenario runs from Precambrian bilaterians or arthropods to the Cambrian arachnomorphs, and then to the trilobites, the arrow of affirmative action in Lin et al. 2006 going from *Parvancorina* to *Skania sundbergi* and then wandering to *Primicaris larvaformis*.[2]

"What is often missed," Matzke argues, "is that deposits like the Chengjiang have dozens and dozens of trilobite-like and arthropod-like organisms." There follows a burst of exuberant thunder: "These are transitional forms!" Matzke is persuaded that whatever is trilobite-lite must be trilobite-like, and so ancestral to the trilobites themselves. The party line is otherwise:

> Early trilobites show all the features of the trilobite group as a whole; there do not seem to be any transitional or ancestral forms showing or combining the features of trilobites with other groups (e.g. early arthropods).
>
> Morphological similarities between trilobites and early arthropod-like creatures such as Spriggina, Parvancorina, and other "trilobitomorphs" of the Ediacaran period of the Precambrian are ambiguous enough to make detailed analysis of their ancestry far from compelling.[3]

If his natural allies in the great cause have refrained from supporting his conclusions, Matzke is prepared to advance them anyway, a policy commanding our admiration, if only for its foolhardiness: "All of this is pretty good evidence," Matzke writes, "for the basic idea that the Cambrian 'Explosion' is really the radiation of simple bilaterian worms into more complex worms, and that this took something like 30 million years just to get to the most primitive forms that are clearly related to one or another living crown 'phyla,' and occurred in many stages, instead of all at once."

This is a view championed by Matzke in defiant isolation. The University of California's Museum of Paleontology makes the obvious case to the contrary:

> When the fossil record is scrutinized closely, it turns out that the fastest growth in the number of major new animal groups took place during the as-yet-unnamed second and third stages (generally known as the Tommotian and Atdabanian stages) of

the early Cambrian, a period of about 13 million years. In that time, the first undoubted fossil annelids, arthropods, brachio-pods, echinoderms, molluscs, onychophorans, poriferans, and priapulids show up in rocks all over the world.[4]

Matzke is pursuing his PhD at the University of California. He is apparently indisposed to visiting museums.

# 12. A One-Man Clade

*"The problems associated with the biological character problem
[cladistics] are so complex and multifaceted and this issue
is so conceptually immature that any single author's account is
doomed to be too narrow and lopsided to be of much use."*
—Günter Wagner[1]

Had Stephen Meyer better appreciated the tools of modern cladistics, Nick Matzke believes, he would not have drawn the conclusions that he did in his book *Darwin's Doubt*, or argued as he had. Meyer is in this regard hardly alone. It would seem that Harvard paleontologist Stephen Jay Gould was just slightly too thick to have appreciated, and the eminent paleontologist James Valentine just slightly too old to have acquired, the methods that Matzke, writing at *Panda's Thumb*, is disposed to champion. Should Valentine be appointed to Matzke's dissertation committee at UC Berkeley, we at the Discovery Institute will be pleased to offer uninterrupted prayers on his behalf. We can offer no assurance of success, of course, but then again, when it comes to cladistic methods, neither can Matzke.

Why, Matzke wonders,[2] did Stephen Meyer not include within his book cladograms such as those he himself displays in his critique, one due to Keynyn Brysse,[3] the other to David Legg?[4] He is, in asking this question, in full Matzke mode: sleek with satisfaction. Meyer may well have refrained from including these cladograms because they are topologically in conflict and display virtually no agreement with one another. Matzke's inability to discern what is directly beneath his nose is hardly evidence of his own competence in cladistic analysis.

It was the German entomologist Willi Hennig who, with the publication of *Grundzüge einer Theorie der phylogenetischen Systematik*, introduced biologists to his scheme of classification.[5] Matzke is surely right to remark that drawing up character sets is a detailed and tedious business. But so is the work involved in alphabetizing the names of all those resident in Moscow in 1937. It is, in either case, no great recommendation. The great merit of cladistic analysis is just the work that it makes for cladistic analysts. Like so much in Darwinian biology, it is a gift that keeps on giving.

A cladistic system expresses a complicated jumble of assumptions and definitions, these expressed most often in the baroque and oddly beautiful vocabulary of Greek and Latin technical terms. No taxonomist with access to words such as *paraphyletic*, *plesiomorphy*, or *synapomorphy* would ever be satisfied by a description of *Anomalocaris* as some bug-eyed monster shrimp. Not me, for sure. Assumptions and definitions in cladistics sheathe a sturdy but simple skeleton, nothing more than a graph, lines connected to points in the plane. The blunt, no-nonsense language of graph theory is quite sufficient. A graph $G = \langle V, E \rangle$ is a collection of vertices and edges. A given vertex may be either an intersection or a terminal point of a graph. A Steiner tree is a graph spanning its terminal points. Although Steiner trees are designed to be unobtrusive, like any skeleton, they make demands all their own, most obviously because they are finite and discrete.

The cladistic classification of the living and the dead proceeds by means of a character matrix, one whose elements are bright but isolated morphological or genetic bits. Fingers are ideal. They can be counted. A person (typically) has no more than five of them per hand. And fingers are discrete. No one need wonder where one begins and the other ends, a point not lost on short-tempered motorists. When character sets are expressed as graphs, the result is a cladogram.

Five taxa in all: A, B, C, D, and E, individuals, species, or whole slobbering groups of them. Assume that this is so. And in 1, 2, 3, and 4, four characters. A four by five matrix suffices to display the distribution of characters, with 1 indicating that a character is present in a taxon, and 0, that it is not:

|   | **1234** |
|---|------|
| **A** | 0001 |
| **B** | 0011 |
| **C** | 1011 |
| **D** | 1111 |
| **E** | 1110 |

The translation of a character matrix into a cladogram can in this simple case be done by hand. Anything more complicated requires a computer program. The clade of cladists and the clade of computer programmers are on the best of terms. One clade washes the other, as professionals so often observe. Terminal points in a cladogram are occupied by the names of taxa, and vertices by their characters.

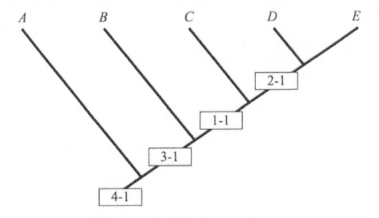

Figure 12.1.

*Aber sei vorsichtig,* as Hennig himself might have warned, and, no doubt, would have had he read Matzke's critique. It is remarkably difficult to read a cladogram without reading something into it, more than the graph conveys, a bit of Darwinian doggerel most often. A, B, C, D, and E are labels marking points in the plane; the taxa that they designate are found in nature. There is a difference. That A is to the left of B is a fact about graphs and labels. It makes no sense to say of two taxa that one is to the left of the other. Very few taxonomists are known widely to confuse their left and their right hands—no more than one or two. This is reassuring. That B is between A and C is otherwise. It is tempting. It is tempting precisely because it invites the taxonomist to undertake an inference from the premise that B is between A and C to the conclusion that B is somehow a descendent of A, an ancestor of C.

A cladogram does not by itself justify anything of the sort: The inference remains a non-starter because it exhibits a non-sequitur. "Evolution[ary theory] is not a necessary assumption of cladistics," the biologist A. V. Z. Brower once remarked.[6] Neither is it sufficient, I would add. A cladogram is one way of depicting the information resident in a character matrix, and given the open-ended relationship between a matrix and its depiction in a cladogram, it is by no means unique. Intermediates that are clear as sunshine in a given cladogram disappear into darkness when the cladogram is rotated.

There is thus

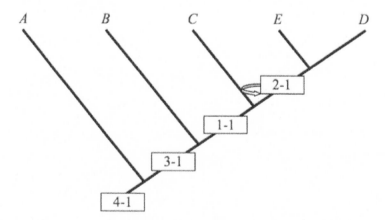

Figure 12.2.

or even this

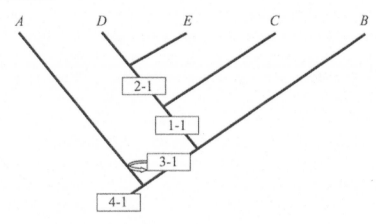

Figure 12.3.

These cladograms preserve the hierarchical structure of Figure 12.1, but they fail notably to keep intermediate taxa where they once belonged. Rotations preserve some of the structure of the original, but not all of it. Similarity in structure may well be an assumption governing the identity of various cladograms; but as these examples show, structural similarity along one dimension does not preserve structural similarity along another. A cladist championing Figure 12.1 in his PhD dissertation is apt to see intermediates in the fossil record that his colleagues, and so his competitors, regard as nothing more than so many graph-theoretic artifacts. Never mind. Championing Figures 12.2 and 12.3 in their dissertations, they have artifacts all their own to exhibit or to hide. (It is a very good thing that these people are seldom armed and rarely dangerous.) If this is so, how, then, to define transitional forms? If no definition is possible, then the relevance of cladistic analysis to Darwinian biology might be more limited than often thought. It is in any case an obvious question to ask, the more so when assessing a book calling attention to the absence of transitional forms in the Cambrian era.[7]

Matzke acknowledges the point without grasping its meaning. "Phylogenetic methods as they exist now," he writes, "can only rigorously detect sister-group relationships, not direct ancestry, and, crucially,... this is neither a significant flaw, nor any sort of challenge to common ancestry, nor any sort of evidence against evolution."[8] But there can be no sisters without parents, and if cladistic analysis cannot detect their now mythical ancestors, it is hard to see what is obtained by calling them sisters. No challenge to common ancestry? Fine. But no support for common ancestry either. Questions of ancestry go beyond every cladistic system of classification, no matter the character states. It follows that questions with respect to the ancestry of various Cambrian phyla cannot be resolved by any cladistic system of classification, however its characters are defined. We are now traveling in all the old familiar circles. The claim made by *Darwin's Doubt* is that with respect to the ancestors of those Cambrian phyla, there is nothing there.

The relationship between cladistics and Darwin's theory of evolution is thus one of independent origin but convergent confusion. "Phylogenetic systematics," the entomologist Michael Schmitt remarks, "relies on the theory of evolution."[9] To the extent that the theory of evolution relies on phylogenetic systematics, the disciplines resemble two biologists dropped from a great height and clutching at one another in mid-air.

Tight fit, major fail.[10]

No wonder that Schmitt is eager to affirm that "phylogenetics does not claim to prove or explain evolution whatsoever."[11] If this is so, a skeptic might be excused for asking what it does prove or might explain.

The graphs and trees of cladistic analysis are, when examined statistically, capable of recording strong signals. Matzke is right to say this.[12] He is himself so alert to them that he resembles an old-fashioned commodore peering beyond mist and mizzen and hoping to see flags. While those signals may be strong, it is often unclear what they are signaling. We may imagine the world's living cladists sorted by a complex character matrix, one involving graduate degrees, publications, tenure, citations, beard length, night sweats, beady eyes, prison records, and trans-gendered identity. Their crown group comprises the living cladists together with their last common ancestor and all of his descendants. Cladistic analysis indicates, I am at liberty to disclose, that Willi Hennig is the last common ancestor of all living cladists, and so an *Ur-Mensch*, another reason for the respect in which he is held. Having branched off early, Sokal and Sneath make for a sister group. Extinct stem groups may be seen as well, tracing their ancestry to an *unter-Mensch*, as German cladists say, one lost in the fog of time and more basal, if not more base, than Hennig himself.

A cladogram of cladists is no different from the real thing, a cladogram by cladists. But Hennig is not the last common ancestor of contemporary cladists, the gorgeous apparatus of cladistics notwithstanding, and no cladistic *unter-Mensch* ever existed. Common characters and common descent are not the same thing.

Cladistic methods thus suggest a number of reservations. Character matrices are the method's heart and soul, ineliminable in practice and theory. It is precisely because character matrices are finite and discrete that cladists believe that they have on hand a body of data that they can master. "Cladistics breaks up the bodyplan characters," as Matzke observes, "and shows the basic steps they evolved in, and also which parts of the 'bodyplan' are actually shared with other phyla."[13] This is precisely what cladistics does, but what it does is at least open to the suspicion that when it comes to these issues, cladistic analysis is driven more by what cladists can do than what they should do. "No experienced naturalist," Stephen Jay Gould remarked, "could ever fully espouse the reductionist belief that all problems of organic form might be answered by dissecting organisms into separate features."[14] By the same token, no experienced linguist would ever claim that the order in which Latin, French, or German words entered the English language explains very much, if anything, about its fundamental structure.

When cladistic analysis is applied to Cambrian paleontology, the imponderables of the method reappear as obscurities in the result, an interesting example of descent with modification. In this regard, Matzke writes, "the arthropods are instructive."[15] And so they are. Let us be instructed. Let us be instructed by Gregory D. Edgecombe of the Department of Paleontology at the Natural History Museum in London. "Arthropod phylogeny," he writes cheerfully, "is sometimes presented as an almost hopeless puzzle wherein all possible competing hypotheses have support." His conclusions hardly amount to a ringing rejection of the hopeless puzzle school, but the best that he can say for his field is that not "anything goes."[16] I am sure this is so.

Matzke is not about to go all hopeless on his supporters either. Witness his discussion of the otherwise hideous *Anomalocaris*. "Anyone actually mildly familiar with modern cladistic work on arthropods and their relatives," Matzke writes, "would realize that *Anomalocaris* falls many branches and many character steps below the arthropod crown group." As it happens, *Anomalocaris* does not fall anywhere: It is the anomalocaridids that do the falling. They in turn are folded within Radiodonta, which makes for an order and comprises a stem-group arthropod. Their most evident character in common with the arthropod crown group is what in a lobster would be called a bony claw. It is not much, but cladists are not fussy. The anomalocaridids include the genera *Anomalocaris*, *Peytoia*, *Schinderhannes*, *Amplectobelua*, and *Hurdia*; the arthropod stem group, the gilled lobopodians, dinocaridids, the taxon incorporating Radiodontia, fuxianhuiids and canadaspidids. It is here that characters drift between Onychophora and the arthropod crown group itself.

Without ever mentioning just which shrimp he has mind, Matzke writes that "it is one of many fossils with transitional morphology *between* the crown-group arthropod phylum, and the next closest living crown group, Onychophora (velvet worms)."[17] With this remark, he solidifies his reputation as a man capable of making the same mistake twice. Common characters? Or at least one in that bony claw? Yes, of course. *Transitional* morphology? *Ah* but no. At best, an intermediate morphology. Nor can *it* be *intermediate* between the crown arthropods and Onychophora. That would be rather like placing a carrot as an intermediate between the United States and Bolivia. Wrong classification. It is Radiodonta as a taxon that is intermediate between taxa.

These are terminological disputes among us experts. A bloop is not necessarily a blunder. Let me refer in what follows to *Anomalocaris X*, where X designates whatever it is that Matzke had in mind. Does *Anomalocaris X* enter the fossil record after the first representatives of the arthropod crown make their appearance? It is in that case, a little late, one might think, to be a transitional form. *Anomalocaris X* could hardly be ancestral to itself nor ancestral to the trilobites and other crown group arthropods. Before then? It is in that case a little too complex to be the ancestor of the trilobites, possessing as do all such bugs compound eyes more sophisticated than anything exhibited by the trilobites—more sophisticated than anything except the eyes of various dragonflies, in fact. What, then, is the ancestor of *Anomalocaris X*? This is just the question that Stephen Meyer asks, again and again, as it happens. It is a part of the Cambrian mystery.

It is with a question such as this that the cladistic method achieves a triumph uniquely its own. We may allow Edgecombe the last word. It is in his Figure 1 that he displays a cladogram for stem and crown group arthropoda. The figure includes Onychophora, which falls outside these stem groups but is nonetheless hoping for cladistic glory. Thick black lines move downward from various stem groups and then stop abruptly where the evidence leaves off. The cladogram nonetheless continues recklessly down through the muck and mist of the early Cambrian, thick black lines now replaced by lines that are thin. These mark the ghost lineages of the cladist's art, the artifacts of his method and not the imperatives of the evidence. While the last common ancestor of Radiodonta is basal to the last common ancestor of the arthropod crown, both are imaginary.

Ghost lineages are often defended, rarely extolled. Like much in cladistic analysis, they represent the withdrawal of a theory from any very robust confrontation with the evidence. They simply cannot be used to attack a view of the Cambrian that begins by questioning whether there is anything behind these ghosts beyond the cladist.

A man who believes in ghost lineages is demonstrably inclined, after all, to believe in ghosts.

# 13. Good as Gould

A MONG OTHER THINGS, MEDIEVAL THINKERS BELIEVED THAT HUMAN beings were unique in ways that were absolute and inviolable. This doctrine many modern biologists have emphatically rejected. "The Western world," Stephen Jay Gould remarks, "has yet to make its peace with Darwin."[1] The great impediment to this reconciliation, he goes on to add in his mad way (the sense strong that he is urging a difficult truth on a dogmatic public), "lies in our unwillingness to accept continuity between ourselves and nature, our ardent search for a criterion to assert our uniqueness." He continues:

> Chimps and gorillas have long been the battleground of our search of uniqueness; for if we could establish an unambiguous distinction—of kind, rather than degree—between ourselves and our closest relatives, we might gain the justification long sought for our cosmic arrogance. The battle shifted long ago from a simple debate about evolution: educated people now accept the evolutionary continuity between human and apes. But we are so tied to our philosophical and religious heritage that we still seek a criterion for a strict division between our abilities and those of chimpanzees.[2]

Now I quote all this not merely because Gould holds a chair at Harvard and I do not, although this made the target all the more tempting, but because Gould represents a charming intelligence corrupted by a shallow system of belief.

No distinction in kind rather than degree between ourselves and the chimps? No distinction? Seriously, folks? Here is a simple operational test: The chimpanzees invariably are the ones *behind* the bars of their cages. There they sit, solemnly munching bananas, searching for lice, aimlessly loping around, baring their gums, waiting for the experiments to begin. No distinction? Chimpanzees cannot read or write; they do not paint, or compose music, or do mathematics; they form no real communities, only loose-knit wandering tribes; they do not dine and cannot cook; there is no record anywhere of their achievements; beyond the superficial, they show little curiosity; they are born, they live, they suffer, and they die.

No distinction? No species in the animal world organizes itself in the complex, dense, difficult fashion that is typical of human societies. There is no such thing as animal *culture*; animals do not compromise and cannot count; there is not a trace in the animal world of virtually any of the powerful and poorly understood powers and properties of the human mind; in all of history, no animal has stood staring at the night sky in baffled and respectful amazement. The chimpanzees are static creatures, solemnly poking for grubs with their sticks, inspecting one another for fleas. No doubt, they are peaceable enough if fed, and looking into their warm brown eyes one can see the signs of a universal biological shriek (a nice maneuver that involves hearing what one sees), but what of it?

One may insist, of course, that all this represents a difference merely of degree. Very well. Only a difference of degree separates man from the Canadian goose. Individuals of both species are capable of entering the air unaided and landing some distance from where they started.

# 14. Ovid in His Exile

Schermerhorn Hall at Columbia University was the scene of many strange experiments. One day, a very young chimpanzee escaped from the building and, flushed with its freedom, began to gambol and frolic on the pathetic square of shabby and well-worn grass that served as a lawn in front of the building. A crowd quickly collected. The mathematician Lipman Bers joined me. A scruffy puppy noticed the commotion and scooted into the square where the chimpanzee was playing. The two animals promptly became friends, but the puppy, it soon became apparent, was less intelligent than the chimpanzee. Again and again he would find himself maneuvered into absurd and humiliating positions. "So stupid," snorted Bers, referring to the dog. Pleased and flattered by the attention, the chimpanzee began to refine his act and play to the crowd, using gestures, and even facial expressions—the universal rictus of triumph, for example—that everyone recognized. After a while, the chimpanzee's frantic owner, a rather dishy young woman, I recall, collared him in the courtyard and the game was over. As the chimpanzee was led away, he waved to the crowd, a true sportsman. The puppy sat on its haunches and panted assiduously.

I learned later from Bers that research biologists were trying to teach the chimpanzee American Sign Language. They had been working with an older animal, but evidently the beast, while learning some signs, grew unsurprisingly to detest his owners, who finally shipped him to a zoo in San Diego. There he occupied himself unprofitably in attempting to teach the other animals to sign, a splendid case of the incompetent endeavoring to instruct the indifferent.

"A vast tragedy," Bers remarked sentimentally, "like Ovid in his exile."

I mention this sad little story only to remark on its ironic conclusion. For a time during the 1970s, a number of biologists were actually convinced that they had taught chimpanzees and great apes to talk; many of them reported long conversations, chiefly about bananas (*Me: More!*), that they held with their charges. Their research was no sooner published than it was accepted and believed, largely, I think, because a crude Darwinian theory—there is no other—made it difficult to imagine that profound and ineradicable differences exist between human beings and the rest of the animal world. Penny Peterson at Stanford, Herbert Terrace at MIT, and David Premack at the University of Pennsylvania all convinced themselves that somehow the great apes had sat in stony silence throughout the vast reaches of biological time only because they lacked human conversational companionship.

The inevitable, skeptical reaction soon set in. Videotapes taken of chimpanzees revealed, when carefully analyzed, that what had passed for chimpanzee conversation was nothing more than prompted signings in the best of cases—a record of the beast's pathetic endeavor to say whatever it was that his trainer wished him to say; in the worst of cases, the beast simply babbled (*More Me More More!*), his signs utterly devoid of meaning. Herbert Terrace, who had wasted years in browbeating the poor creatures, examined videotapes of his own encounters with his animals and came away shaken. Some work, of course, continues, but to little effect. Ever credulous, scientists now report that they have engaged the dolphin in stimulating conversation. Next year, no doubt, it will be the turkey.

Seventeenth-century Jesuits wondered why dogs do not talk. Their conclusion bears repeating. They have nothing to say.

# IV. Past, Present, Future

# 15. A Natural History
# of Curiosity

## Ulugh Beg, Al-Ghazali, and Taqi al-Din

During the living centuries of the Arab empire, a series of stellar observatories glittered like jewels throughout the archipelago of its conquests. The observatory played an important role in the religious life of devout Muslims. More so than either Jews or Christians, men of the faith were called upon carefully to mark the schedule of their devotions. Caliphs in Baghdad counted time by means of either a water clock or an hourglass, and yet the *Quran* commanded five-fold prayers each day, and it commanded the faithful to face the shrine of *Kaaba* in Mecca as they prayed—tasks requiring some very considerable mental dexterity. "At the last Judgment," the Turkish devout Said Nursî remarked, "the ink spent by the scholars is equal to the blood of martyrs."[1]

Those scholars celebrated at the last judgment were apt to be scholars of religion and so bound by the inerrancy of the *Quran*. "Allah turns over the night and the day," reads a well-known Quranic verse, "most surely there is a lesson in this for those who have sight." It is hardly surprising that Muslim mathematicians and astronomers, from the late seventh to the early fifteenth century, regarded their curiosity, on those occasions when they were called upon to justify it, as if its indulgence were an exercise calculated to increase their devotion.

But of all the human emotions, curiosity is the one least subject to the general proscription against gluttony, and once engaged, even if engaged initially in the service of religion, it has a tendency to grow relentlessly until in the end the scholar becomes curious about the nature of revelation itself.

## Ulugh Beg

Muhammad Taragai Ulugh Beg was born in 1394, and died fifty-five years later, the victim of an assassination orchestrated by his son. The grandson of the murderous Tamerlane, Ulugh Beg became the ruler of Transoxiana on the death of his father. It was in Samarkand that he created an outstanding astronomical observatory.

A king and an astronomer, Ulugh Beg's moral nature impressed his contemporaries. He was admired. Writing to his father, a young man by the name of Giyâth al-Din Jamshid al-Kâshi was concerned to establish Ulugh Beg's reputation as a scientist as well as a ruler. "The King of Islam," he writes, "the issuer of orders to the seven climes, may God preserve his realm and sovereignty, is a learned person." There follows an earnest disclaimer. "I do not write this and make these assertions out of politeness." Various encomia now follow. Ulugh Beg's knowledge of the *Quran* is impeccable and wide ranging. He knows most of the *Quran* by heart; he recites at least two sections before experts each morning. He makes no mistakes. His knowledge of grammar and syntax is very good, and he writes Arabic very well. Ulugh Beg, Al-Kâshi assures his father, is well versed in jurisprudence, logic, and the theory of literary styles.

If Ulugh Beg emerges from Al-Kâshi's letter as a man of parts, as he surely does, it is his scientific stewardship that most elicits Al-Kâshi's admiration. Beyond his competence, there is his temper: rare in a scientist, unheard of in a king. According to Al-Kâshi, Ulugh Beg was determined to act in his observatory as one scientist among others:

If in certain cases, there happens to be anything concerning which we, his servants, have some doubt, the point is discussed, and no matter from what side the clarification of the mistake comes, His Majesty will at once accept it without the least hesitation, [for] it is his aim to see that everything is thoroughly investigated and to have the work at the observatory accomplished in the best possible manner.[2]

Al-Kâshi died as a young man, his death leaving his colleagues stunned and mournful. When a few years later, Ulugh Beg composed the introduction to his monumental *Zij*—the observatory's astronomical tables—he reflected on his own role in the work of the observatory, and supplied as an aphorism the noble living voice that is missing from Al-Kâshi's letter: "Our accomplishments indicate what we are; look therefore at the things we have left behind." He then credited his teachers and masters, "the most learned of the men of learning," and finally with his own voice now in its proper register, he wrote in tribute to his friend, "the pride of the sagacious people of the world... the unraveler of the intricacies of problems... Giyâth al-Din Jamshid al-Kâshi, may God refresh his resting place."[3]

The observatory that Ulugh Beg created was conceived on a monumental scale. Muslim astronomers quite understood that celestial phenomena are periodic, the stars returning roughly to their position after days, months, or years; but they understood experimental error as well, and they knew that careful observations could be defeated by a cloudy night. The observatory at Samarkand was dedicated to a thirty-year program, one longer than anything envisioned, by way of comparison, for the Hubble space telescope. The observatory was grand in its physical aspects. The sextant used to measure the ecliptic had a radius of almost forty meters, roughly comparable to the radius of the reflecting telescope at Mt. Palomar. The building was constructed of marble, and it housed exquisite brass and copper instruments. After Ulugh Beg's death, the observatory was sacked by local rulers, unable to resist its wealth, and by local clerics, unwilling to abide its learning.

By 1449, the observatory was in ruins.

The ancient Near East is filled with the debris of centuries. The physical glories of the observatory at Samarkand are incidental. It is the moral instrument created by Ulugh Beg that is compelling, for that instrument, in various copies and clones, survives into the twenty-first century. The most perceptive contemporary scholar of the Islamic observatory, Aydin Sayili, assigns to the institution the power of primogeniture: "The observatory as an organized and specialized institution was born in Islam; [and] it passed on a rather highly developed state to Europe, and this was followed shortly afterwards, by the creation of modern observatories in Europe, in an unbroken process of evolution."[4]

Writing in 1420 or 1430, Ulugh Beg described science in a way that suggests nothing of the martyr's blood. "Intellects are in agreement," he wrote, "and minds are in accord as to the excellence of science and the worthiness of scientists."[5] The benefits conferred are very often matters of self-improvement. "Science sharpens the intellect and strengthens it; it increases sagacity, and augments perspicacity."[6] Benefits are social as well as personal. Those sciences whose principles are "indisputable and self-evident" have the merit of being "common to people of different religions."[7]

Is there any reason to think any of this true? It would be nice to think so.

At the beginning of the twelfth century, the Arab archipelago stretched from Cordoba in the west to the border of India in the east. The greatest of the medieval caliphs, Harun al-Rashid, had, in the early ninth century, created the *Bayt-al Hikmah* (the House of Wisdom); his son, Al-Mamun, invested the resources of his empire in its efflorescence, and until Mongol primitives overran Baghdad in the thirteenth century, the *Bayt-al Hikmah* retained the light of its learning, flickering wherever mathematicians, logicians, translators, goldsmiths, physicists, physicians, astronomers, astrologers, or poets gathered, whether in fly-specked provincial villages, the coffee house a place of culture, camels neighing irritably on the streets, the men gabbling in Uzbek or in

Turkish, or if not there, then in Baghdad, where pleasure boats plied the Tigris and colored lanterns decorated all the stately riverfront mansions.

By the eleventh century, the *Bayt-al Hikmah* had become, like the library at Alexandria, an institution beyond space and time. The most notable physical scientist of the eleventh century, Abu al-Biruni, was born in Uzbekistan. A man of overpowering curiosity, his collected works run to thirteen thousand folio pages, covering mathematics, astronomy, physics, linguistics, and a dozen other subjects. He wrote an extraordinary account of Indian history, the *Kitab al-Hind*, one in which he observed with some equanimity that among Hindus, Muslims were thought low, impure, cunning, grasping, and cruel.

That Allah is omniscient, he observed with some asperity, does not justify ignorance.

## Abu Hamid Muhammad al-Ghazali

The most perceptive of the Arabic philosophers, Abu Hamid Muhammad al-Ghazali, was born in Baghdad in 1058. If his portraits are inauthentic or unverifiable, they are remarkable in that not one of them depicts a man with a merry face. He seems to have emerged from the womb with burning eyes. Obsessed in childhood by the divine, he possessed in adolescence a mature grasp of dogma and doctrine. He was trained in the law and in theology, and was widely considered matchless in debate.

A curious and incomplete kind of elective affinity passes from the Muslim to the Christian world in the high Middle Ages. The beetle-browed Al-Ghazali is a coeval of sorts to Peter Abelard, who was born in 1079 in what is now Brittany. Leaving home at the age of fourteen, he was introduced to eleventh-century philosophy and disputation by the philosopher Jean Roscelin. Abelard then wandered the Loire valley, "disputing," as he says, "like a true peripatetic philosopher, whenever I heard there was keen interest in the art of dialectic."

He was very widely considered insufferable.

Abelard entitled the book that prompted the nerve of theological heresy to commence vibrating *Sic et Non* (Yes and No). "There are many seeming contradictions," Abelard wrote, "in the writings of the Church fathers." The theologian must therefore address the church fathers with what Abelard called "the full freedom to criticize and with no obligation to accept unquestioningly."

Bernard of Clairvaux, the largest religious personality of the twelfth century, denounced Abelard as a "scrutinizer of majesty and fabricator of heresies."[8]

Abelard was a logician by profession, and at forty he came to love late in life, the coldness of his craft doing little to calm his boiling blood and nothing whatsoever to improve his better judgment. Long after he had lost his love, along with his manhood, drunkards in all the low Parisian taverns were still banging their tankards on the tabletops and singing the songs that Abelard seems effortlessly to have composed. Not one survives, although some vagrant ditties attributed to Héloïse may still be found in Breton folklore: *mon clerc, mon bien cher Abailard.*

At some time during the late eleventh century, Al-Ghazali suffered a terrifying spiritual crisis. He became overwhelmed by doubt. Skepticism is today an honorific. Men who believe any number of absurdities are pleased to regard themselves as skeptics and gather unpleasantly in conventions to say that this is so. The late Christopher Hitchens was in this respect masterful. What can be asserted without evidence, he observed sonorously, can be dismissed without evidence. The remark has been promoted to Hitchens's Razor on Wikipedia. Hitchens never once considered that Hitchens's Razor might be used to shave itself. "It is just a sentence," he said amiably when asked.

This is not the kind of skepticism from which Al-Ghazali suffered. His sense of doubt was corrosive; it was annihilating. He could find no reason for belief in the sciences at his command, and he came to distrust his senses. Unable to combat doubt and unwilling to accept it, Al-Ghazali stood naked in his torments. God put a lock on his tongue. It is a source of regret that the deity is presently not inclined further to distribute locks. His physician, a smooth Baghdad professional, re-marked that there could be no way to treat his affliction unless "his heart [were] eased of its anxiety."

Al-Ghazali's most profound skeptical arguments he expressed in a tract entitled the *Tahafut al-Falasifa* (the Incoherence of the Philoso-phers). It is a bitter work and one written before his spiritual crisis. Have the philosophers maintained that the world is the necessary effect of some divine and so necessary cause? Apparently they have. Does God then act as He must or as He would? If He acts as He must, He is not God, and if He acts as He would, He is not necessary.

The most famous of Al-Ghazali's arguments in the *Tahafut* con-cerns the relationship between causes and their effects. "The connection between what is habitually believed to be a cause," Al-Ghazali wrote, "and what is habitually believed to be an effect, is not necessary accord-ing to us." There is no necessary connection between "the quenching of thirst and drinking, satiety and eating, burning and contact with fire, light and the appearance of the sun, death and decapitation, healing and the drinking of medicine, the purging of the bowels and the using of a purgative, and so on to [include] all [that is] observable among connected things in medicine, astronomy, arts and crafts."[9]

Seven centuries later, the suave Scottish philosopher David Hume advanced a similar argument. If A causes B, then either something or nothing makes A the cause of B. If something, what is it, and if noth-ing, why is A the cause of B? That there was something, Hume never doubted; but what it was, he could not say. The "ultimate springs and principles" of nature, he concluded bleakly, "are totally shut up from hu-man curiosity and enquiry." Hume appealed to custom to explain what he could not directly discern: "We assert that, after the constant con-junction of two objects—heat and flame, for instance, weight and so-lidity—we are determined by custom alone to expect the one from the appearance of the other."[10]

This is no very persuasive example of analysis. A is said to cause B when the two always occur together. If A is the cause of B, then A and B cannot occur *together*, and neither A nor B can occur *again* or could have occurred *before*. They are what they are; they took place when they did. Far from explaining that A causes B, custom makes reference only to events that are *like* A and *like* B. If we are unable to analyze why A caused B, an appeal to similar circumstances is unavailing. Ten drowning men are unlikely to be of support to a man who is drowning.

This Al-Ghazali quite understood. Only some principle of binding necessity makes causation coherent. No such principle seemed evident to him, and no such principle seems evident to us either. Al-Ghazali settled his accounts with the concept of causation by appealing to the whim of God.

Neither Al-Ghazali nor Hume settled their doubts to their own satisfaction. Were those doubts ever real? What would have satisfied either man if asked to offer an imaginative account of the connection between A and B? There is nonetheless a difference in attitude between Al-Ghazali and Hume. The secret springs that Hume invoked, although they could not be described, Hume nonetheless thought real. Perhaps there is *nothing* in nature and so *no* causal connection between events. Things just happen. Chaos predominates. This is not an idea that Hume entertained; it is very tempting to suppose that Al-Ghazali did. It is the difference between them.

## Taqi al-Din

Lying undisturbed in the locked drawer of the great library at Istanbul is a manuscript entitled the *Shahin-Shah-nama* (Book of the King of Kings). Its author was a Persian, Ala ad-Din Mansur-Shirazi. Written in verse, Mansur-Shirazi's poem is in fact a chronicle, one describing the reign of Sultan Murad III, the ruler of the Ottoman Turks from 1574 to 1595. The poem itself, Mansur-Shirazi piously comments, was completed on October 28, 1581, the last day of Ramadan. The manuscript, its translator notes, is "richly illustrated and illuminated," in the ornate Persian tradition—no perspective, the turbaned characters in the illustrations clambering up the folio pages as if they were climbing so many beanstalks. The binding, he adds, is "quite elaborate," but the penmanship, considered from "the artistic point of view," is disappointing.

Parts of the poem, comprising 153 folios, are purely formulaic. God is praised, the Prophet thanked. There are various prayers. But the greater part of the poem is historical, and among the events that it records, there is the story of the creation and destruction of the Royal Istanbul Observatory, the successor by 150 years of the observatory at Samarkand.

Mansur-Shirazi begins his poem by discussing the Royal Observatory's instruments, remarking that "our wise star gazing astronomer" has placed the observatory in Istanbul on such a firm foundation that in the current era—the late sixteenth century—the prestige accorded astronomy is comparable to the prestige once accorded "the science of religion." It is an odd remark, and one intended to excite apprehension. The preparations undertaken by Taqi al-Din, the director of the observatory, are recounted in some detail: "The required instruments all became ready/ They were, with their brass and copper sections, of great perfection."

A number of verses that follow are given over to hopes that are by turns exuberant and pious. The appearance of a comet during the month of Ramadan in 985 is recounted, its implications discussed. Until this point, the poem reveals nothing of note. Things now change dramatically. Taqi al-Din is summoned to the presence of Sultan Murad III. A strange concourse follows. The Sultan addresses Taqi al-Din as "you witty man of conscientiousness and perfection," but the flattery, in addition to being obscure, is insincere. "People of learning," the Sultan remarks—and it is now clear that he means clerics—have made inquiries. They have expressed reservations. And just what *has* the observatory been doing?

Taqi al-Din answers as almost any physical scientist would, although with a certain purely Turkish flourish: "In the *Zij* of Ulugh Beg, there were many doubtful points; now through observation, the tables have been corrected, and out of grief, the heart of the foe has writhed and twisted in its coils."

It is now that history breaks in two. The Sultan issues a shocking order, one that the poem has done nothing to suggest. He summons his Admiral and commands that he should promptly "wreck the observatory, and pull it down from apogee to perigee."

And it was done: "Nothing remains of the observatory but a name and a memory."

The reader might well expect the poet to share in his consternation, but far from castigating the acts that he records, he denies his own emotional sentiments and provides a melancholy justification for the Sultan's actions: "In the labyrinth of this existence of short duration do not close your eye of circumspection out of greediness and appetite. Do not make decisions concerning the affairs of the firmament."

The justification given is then elaborated in what would appear to be a direct address to men of science:

… What hope then can you have of uncovering these matters
That you make diversions from the surface of the earth
And indulge in celestial affairs.
Come, let us get away from this egotism and wrangling
For the old decrepit world is monstrously tricky and deceptive
Beware of her for she may put our affairs in confusion.

This is hardly advice that a partisan of inquiry would wish to hear. It is nonetheless the advice—and the only advice—that Mansur-Shirazi is prepared to offer:

When the affair concerning the Observatory was brought
   to completion
And it was torn from its foundation and its traces were
obliterated
All people of faith prayed for the mighty King
For he had caused the performance of a deed
Which was in accordance with the Law of the True Religion.

The year is 1580. Just five years earlier, Tycho Brahe had laid the foundations of his own observatory at Uraniborg, the enchantment of seeing dying in one culture, even as it was coming to life in another.

# 16. The Ineffable Higgs

Surely its discovery meant something? The Standard Model (SM) of particle physics demanded its existence, after all; and the demand was met. If it took forty years and more than sixteen billion dollars to discover the thing, physicists could with satisfaction observe that the public got what it paid for, the first step, of course, in demanding that the public pay for more of what it got. Photographs of Peter Higgs staring tenderly into space, *ses yeux perdus*, conveyed an impression of appropriate intellectual satisfaction.

The discovery was announced; the story reported; and then there was silence. Physicists endeavored, of course, to maintain the impression that they had discovered something of inestimable value. From the headline of a *Daily Beast* article by Sean Carroll: "The Higgs boson discovery revolutionizes the world of physics."[1] If this is what physicists always say, then, at least, they seem never weary of saying it.

Lawrence Krauss, writing in *The Daily Beast* as well, gave it his best. Many years ago, Leon Lederman had designated the Higgs boson as the God particle. No one can today remember why. The *God* particle? "Nothing could be further from the truth," Krauss remarked.[2] In this, of course, he was entirely correct: Nothing *could* be further from the truth.

In the end, Krauss, like Carroll before him, could do no better than an appeal to the revolution. The discovery of the Higgs boson "validates an unprecedented revolution in our understanding of fundamental physics." Readers of *The Daily Beast* are always pleased to uphold the revolution, no matter how revolting. Yet, the Standard Model was completed in the early 1970s, the revolution it conveyed having begun in the late 1920s, circumstances that might suggest the gradual emergence of a soberly modulated consensus more than anything else—the perfect truth, as it happens.

If the revolution is either far away or long ago, there is always God. The discovery of the Higgs boson does nothing to confirm his existence, Krauss argued, therefore it must do everything to diminish his relevance. And so it does. The Higgs boson, he wrote, brings "science closer to dispensing with the need for any supernatural shenanigans all the way back to the beginning of the universe—and perhaps even before the beginning, if there was a before."[3]

About this declaration, since it countenances a *before* before a beginning or a beginning before a before, all that one can say is that Krauss has covered his bases.

§

The Standard Model (SM) of particle physics was created to impose order on elementary particles that during the 1960s had seemed to multiply in proportion to the funding available to determine them. If there were very many particles before the advent of the SM, there remained very many particles after its advent—reason enough to wonder whether any of them were really *elementary*.

Whatever their ultimate nature, the elementary particles are today divided into fermions, hadrons, and bosons. Fermions come in twelve flavors arranged in two families of leptons and quarks. The quarks form a three-fold family structure of their own: up or down, charmed or strange, and at the top or loitering in the bottom. These twelve particles give rise to twelve anti-particles.

The bosons comprise the photon, the gluon, and the electroweak bosons, of which the Higgs boson is now king.

And the hadrons are organized as families of bound quarks, and appear either as baryons, which are composed of three quarks, or as mesons in which quarks cohabit amicably with anti-quarks.

The SM has many virtues: simplicity is not among them.

§

The SM is both a scheme of classification and a system of explanation, one encompassing quantum electrodynamics, and theories of the weak and strong nuclear forces. These theories are not by any means the same, but they are expressed in the common language of quantum field theory, and it is quantum field theory that the SM serves to make great.

Quantum mechanics was created in the 1920s to explain experimental results that defied classical explanations. Given the choice of passing through one of two experimental slits, a single photon somehow managed to pass through them both. The photon was thus a wave. Other experiments had before suggested that it was a particle. Quantum mechanics accommodated experience by defying common sense. The photon was both a wave *and* a particle.

Quantum mechanics and Einstein's theory of special relativity, physicists understood, were not obviously in conflict, but neither were they the best of friends. Special relativity drew a tight connection between energy and mass; and quantum mechanics was professionally engaged in dissolving tight connections.

With respect to a particle's position and momentum, the axis on which the old world turned, classical physics had been absolute. The physical properties of a particle were accessible to measurement and were accessible to measurement all the way down.

In quantum mechanics it is otherwise. In 1926, Werner Heisenberg argued that the formalism of quantum mechanics placed limits on measurement. The more certain a particle's position, the less certain its momentum. Measuring them both to the same level of accuracy was impossible and it was impossible in principle—reason enough to suppose that the uncertainty principle was a principle of nature, and not an artifact of measurement.

The uncertainty principle and special relativity assigned to an otherwise empty region of space the power to produce new particles from the void. The greater the uncertainty about the void, the more it seethed, and since at short distances, it did a great deal of seething, it was at short distances that it seethed to most productive effect.

Any version of quantum mechanics incorporating special relativity would necessarily be a theory of many particles. They were potentially all over the place.

And for this, a field was required. It was this that Paul Dirac provided. The date is 1927.

§

Quantum mechanics broke the distinction between particles and waves, quantum field theory the distinction between particles and fields. A field is something like the wind: it is everywhere in space and nowhere in particular. Within its ambit, particles appear as focused ripples, knots of a sort, some temporary but countable tightening of things.

Quantum field theories were by no means an immediate success because they were by no means consistent with themselves. Whenever they were applied, infinite magnitudes appeared, glum, glowering, irremovable. New mathematical techniques were required to remove them, and although they worked, physicists did not know why they worked and used them with the awkward sense that for every good reason, they were doing a very bad thing.

Quantum electrodynamics was completed in the late 1940s by Richard Feynman, Julian Schwinger, and Sin-Itiro Tomonaga. Theoretical calculations of the electron's magnetic moment agreed with experiments to one part in a billion. It is an agreement that may fairly be described as freakish. Having sinned to such good effect in the 1940s, physicists continued to sin thereafter. In the 1950s, Chen Yang and Robert Mills proposed a daring generalization of quantum electrodynamics; and in the 1960s, Steven Weinberg, Sheldon Glashow, and Abdus Salaam took the Yang-Mills equations and ran like the wind. They stopped when they had a theory of the weak nuclear force in hand.

And more—far more. The idea that there is a form of unity beneath the diversity of experience is one that has animated the imagination of physicists since the seventeenth century. It is a powerful but not an obvious idea. One of the achievements of the SM is the demonstration that at high energies, the weak and the electromagnetic force are the same. The W (W+ & W–) and Z bosons are the vehicles by which the weak nuclear force is mediated. They are nicely described in terms of a theory that establishes their underlying symmetry. Theoretical considerations suggested—they *demanded*—that, like the photon, these particles should have no mass. The W and Z bosons were massive: they were simply huge. In the work that earned them their Nobel Prize, Weinberg, Salaam, and Glashow argued that the symmetry controlling these bosons must have been broken. It is the Higgs mechanism that accounts for the symmetry breaking and so accounts for their mass.

There remained the strong force and its interactions. Although quarks were required in theory, they were impossible to observe in experiment. It was not until David Gross, David Politzer, and Frank Wilczek proposed their theory of quark confinement that physicists were able to affirm that what could not be observed should not be observed because it could not be observed. Although Weinberg, Glashow, and Salaam had demonstrated a form of unity between the electromagnetic and the weak force, the strong force remained like Achilles, defiant in its isolation.

The SM accommodates three forces and it expresses three theories. It is obviously incomplete. It does not encompass the force of gravity. What it does, it does very well, but if tomorrow it would require renovation, no physicist would worry overmuch. Theories come and go. Quantum field theory is otherwise, a way of thought, irreplaceable. Among particle physicists, the conviction is almost universal that quantum field theory is the structure chosen by nature to conduct her affairs.

"Absolutely all phenomena," Ptolemy wrote in a slightly different context, "are in contradiction to the alternate notions that have been propounded."[4]

§

And this raises a question at least as difficult as any raised by particle physics itself. Just how should claims of this sort be understood and, if understood, assessed? It makes no sense to take particle physicists at their word. It is their word that we are uncertain about taking.

There is in every human being something like a vacuum state, a lowest energy level in which physical judgments are spontaneously made, a zone of comfort. Evolutionary psychologists often describe this vacuum state as folk physics. I would prefer to see *die Völker* left in *lederhosen*, but the idea is useful because it is inescapable. No one, of course, remotely suggests that the SM is much to be admired because it is *gemütlich*. Popular explanations are plainly absurd. "Here's the secret of the Higgs mechanism," Sean Carroll observed in *Discover*: "when you spontaneously break a gauge symmetry, the would-be Nambu-Goldstone boson gets 'eaten' by the gauge bosons!"[5] The metaphor by which one boson eats another, although widely employed, means nothing whatsoever.

What is required of quantum field theory is not its translation into common sense. The physical sciences have been in retreat from common sense since the seventeenth century. There is nonetheless a distinction as important in the sciences as it is in military affairs. It is the distinction between a retreat and a rout. If the SM and quantum field theory are to have a claim on our intellectual allegiance, there must be some scheme by which the ideas that they embody may be recovered in a way that makes simple, persuasive, and intuitive sense. It is not enough in this respect to say that certain ideas work because the predictions they make possible are accurate. That goes without saying. The question is *why* this is so.

It is in mathematics and not theoretical physics that judgments of this kind are made and then enforced. Across the vast range of arguments offered, assessed, embraced, deferred, delayed, or defeated, it is *only* within mathematics that arguments achieve the power to compel allegiance because they are seen to command assent. And it is only by means of mathematics that the powerful ideas of an alien discipline such as theoretical physics may step by step be returned to the ordinary human power to grasp things without mediation and so to grasp things at once.

Quantum field theories are, of course, expressed in mathematical language. What else is there? But it is one thing to use a language, quite another to accept its discipline. Newtonian mechanics has in the twentieth century been formulated in exquisite mathematical detail. If Newton would not have understood a word of it, that is evidence only that physics and mathematics have different aims. What gives pause in the case of quantum field theory is just the odd fact that when quantum field theory *is* expressed in terms of the ancient strictures of mathematical rigor, the result is confusing.

§

Until the 1960s, particle physicists accepted the division of things into particles and fields. Which came first? In a series of superb lectures delivered during the 1970s, Richard Feynman was unequivocal. A *particle*, he said, *take my word for it*. Particle accelerators are, after all, in the business of accelerating *particles*. It followed the particles were primary. The fields came afterwards.

The contrary view is equally compelling. "Since radiation is made only of fields," the physicist Art Hobson observed, "it would be surprising if matter were made of particles. Why should the universe be made of two such different building blocks?"[6] It is fields that come first. The particles can look after themselves. "Particles are epiphenomena arising from real fields."[7]

This is a view endorsed by Steven Weinberg: It is the majority view. "In its mature form," Weinberg writes, "the idea of quantum field theory is that quantum fields are the basic ingredients of the universe, and particles are just bundles of energy and momentum of the fields.... Quantum field theory hence led to a more unified view of nature than the old dualistic interpretation in terms of both fields and particles."[8]

Nature evidently detests a dualism.

What is curious about the confidence with which these views are expressed is just the fact that they have defied the best efforts of mathematicians to make good sense of them.

Some difficulties are of very long standing, so much so that theoretical physicists feel free to ignore them. A theorem published by Rudolf Haag more than sixty years ago and generalized promptly by A. S. Wightman and D. Hall demonstrated that to the extent that either free or interacting fields are wandering in space, there is no one space into which they may expand. A choice must be made among them. No one knows how the choice should be made.

Mathematicians have endeavored to create something like an axiomatic quantum field theory since the late 1940s. Their theories, where they have been successful, have not been complete, and where they have been complete, they have not been successful. Mathematicians have not been able to capture lucidly the idea of a quantum field itself. The ontology of axiomatic quantum field theory is very different. It comprises an axiomatically specified net or a collection of abstract objects known as operator algebras, and these contain the requisite information about the observable elements of a given theory.

The details are interesting, but not relevant. Axiomatic quantum field theory and quantum field theory are divided in their interpretation of reality.

"Quantum foundations are still unsettled," the melancholy Hobson remarks, "with harmful effects on science and society."[9]

This is half false. Quantum physics vexes no man. If half false, then half true.

No one quite knows why mathematicians have been unable to settle even the simplest of questions about quantum field theory: What are the fields *about?*

And with this question, another, more insistent, and, indeed, more fundamental: Why are the methods of quantum field theory so astonishingly successful?

§

Questions of this sort ramify through particle physics. The experiments designed to confirm the theory of electroweak unification were bound to one state of affairs in a particle accelerator; that state of affairs was in turn interpreted to reflect the same state of affairs of the early universe. Something quite new in the history of physical thought was involved. It is something for which entirely appropriate standards of assessment are still lacking.

The chief aim of the experiments confirming electroweak symmetry was to discover conditions under which that symmetry was manifest, and not latent. The conclusion of the experiment was that those conditions, if they were present in the accelerator, were also present in the early universe. If the early universe is unrecoverable, particle accelerators represent an extraordinary anomaly in the ordinary conditions of matter. In the world as we find it, there is nothing like the electroweak symmetry. If these experiments are idealizations, of what are they idealizations?

And what is the connection between these experiments and experience itself?

There are only two possible answers. Either the connection is metaphysical, or there is none at all. Physicists such as Krauss almost automatically offer a metaphysical explanation. "The properties of matter," he writes, "and the forces that govern our existence" are derived from the interaction among fields.

There is in this quotation a three-fold concourse between the Higgs field, the properties of matter, and the forces that govern our existence. The concourse must follow an inference: Given the SM, the properties of matter and the forces that govern our existence follow.

*There is, of course, no way in which to test this inference experimentally.* What conceivable experiment could coordinate matter in an immensely anomalous state within a particle accelerator and the ordinary facts of life?

If no experiment can do what cannot be done, there remains only standards of consistency and coherence, and since no one has ever assigned a clear meaning to coherence, there remains only consistency.

But consistency is a weak constraint. It is weak in the sense that consistent theories need not be satisfied in a single model or universe.

The conclusion that follows from these observations is inevitable. There is no such thing as *the* world, or *the* universe to which the SM unequivocally points. And so there can be no large and general conclusions about what the study of the world, the universe, or Nature reveals about the existence of God.

The image appropriate to physics in the twentieth century is Bruegel's *Tower of Babel*. That the thing is incoherent, everyone can see. What is rarely noticed is that it remains standing.

This is its great achievement.

Figure 16.1.
Peter Bruegel the Elder's *Tower of Babel*, 1563—oil on wood panel.

# 17. The Good Soldier

*"Vanity of vanities, saith the Preacher, vanity of vanities; all is vanity."*[1]

UNTIL THE END OF TIME by BRIAN GREENE
PENGUIN RANDOM HOUSE, 448 PP.

BRIAN GREENE HAS BEEN BUSY AS A BEE. SINCE THE PUBLICATION of *The Elegant Universe* in 1999, Greene has published *The Fabric of the Cosmos, Icarus at the Edge of Time, The Hidden Fabric of Reality, Incidents of Travel in the Multiverse,* and *Light Fails.* Readers have enjoyed his company. And for every good reason. Greene is clever, chatty, accessible. If readers have enjoyed his company, physicists have welcomed his support. They can depend on him. He is a good soldier.

Greene has now published a new book. *Until the End of Time* is an account of how the universe began and how it might end. Things began with a bang and they will end in a whimper. The second law of thermodynamics prevails. The bang is fine, as generally bangs are, but no one can read about the End Times of the universe feeling spiritually refreshed. "As our trek across time will make clear, life is likely transient, and all understanding that arose with its emergence will almost certainly dissolve with its conclusion. Nothing is permanent. Nothing is absolute."[2]

Given that nothing means anything in the long run, Greene wishes to know what it all means. He is unwilling to say that it does not mean a thing, and unable to say why it does not. The multiverse offers him only its most exiguous consolations.[3] On clambering to the edge of space and time, Greene expects to find his double, and so someone exhibiting the same deplorable eagerness to seek relief in travel as Greene himself.

It is no wonder that both men are depressed. Greene and that infernal double of his are not alone. So far as the declension of bleakitude goes—*bleak, bleaker, take-out, Twitter*—they have me at their side, a one-man multitude. At more than four hundred pages, *Until the End of Time* is very long and some readers—not me, of course—may feel on reaching its end that they have understood its title in a personal way.

## A Foundational Fissure

Greene is a distinguished theoretical physicist, well known for his work on mirror symmetries. The laws of physics are there when he wants them. Whether they are there when he needs them is quite another question. "By fully grasping the behavior of the universe's fundamental ingredients," Greene writes, "we tell a rigorous and self-contained story of reality."[4] We do? And rigorous, too? Thermodynamics itself is not obviously derivable either from Newtonian or quantum mechanics, and, if derivable, then not rigorously derivable.[5] Never mind. Rigor is just the party line and party stories are like that. The party's eschatology remains battle worthy. "I can envisage a future," Greene writes, "when scientists will be able to provide a mathematically complex articulation of the fundamental microphysical processes underlying anything that happens, anywhere and anywhen."[6] Sweet is sweet, Democritus observed in lines that Greene quotes, "bitter is bitter, hot is hot, cold is cold, color is color; but in truth, there is nothing in Nature but atoms and the void."[7]

Whatever that far future, in the heat of battle Greene finds himself forced to appeal to the laws of physics *and* Darwin's theory of evolution.[8] Something like Darwin's theory is necessary in order to accommodate the facts of life: its existence and emergence, consciousness, language, freedom of the will, religion and religious experience, meaning. This double dependency introduces a fissure into the foundations of Greene's argument. Darwin's theory is its own best man, and, as often, its own best friend, but the theory does not follow from any "mathematically complex articulation of the fundamental microphysical processes." The theory of evolution and the Standard Model of particle physics may both be satisfied in the real world, but Greene requires a scheme in which the theory of evolution emerges inferentially from "the fundamental microphysical processes underlying anything that happens, anywhere and anywhen." This he does not have. And for the most obvious of reasons. There is no such scheme.[9]

On those occasions in which Greene does not have the law on his side, he pounds the table. "Make no mistake," he thunders, "we *are* all bags of particles."[10] Whereupon there is an objection. The witness is being led, and by the nose, too. It is one thing to open a bag and see that it contains cookies; quite another, to open a human being and see that it contains a variety of bosons. If no one has performed the requisite experiment, it is because no one knows where to look. Theoretical physics is of no help. Michael Peskin and Daniel Schroeder's *An Introduction to Quantum Field Theory* says nothing about human beings, still less about the claim that they are bags of particles.[11] The conclusion that Greene affirms appears, if it appears at all, only at the end of a very long inferential chain, one whose links are missing, or incomplete, or conjectural, or badly soldered, or on the point of breaking.

Your witness.

## Armies of Unalterable Law

Albert the Great explained the stability of the universe by an appeal to Christian theology. *Deus est ubique conservans mundum.*[12] God is everywhere conserving the world. Although fastidiously disguised, this is an idea that reappears in *Until the End of Time*, Greene assigning to the laws of physics the presumptive power of the Holy Ghost. On page 118, Greene remarks that "everything emerges from the same collection of ingredients, *governed* by the same physical principles;" and on page 148, that his "focus is on the existence of laws that *govern* your next move." One paragraph later, they are at it again, those laws, *governing* inanimate collections," and with a little more body English on Greene's part, "*total control*" appears within their grasp. The laws of nature are both "immune to all flattery" and "incapable of loosening the reins."[13] A bit later, Schrödinger's equation is found "deterministically chisel[ing]" its intentions on stone. Imagine what it might chisel if given access to Mt. Rushmore.

This is a very old way of thinking, full of an unwholesome allure.[14] Philosophers since Hume have understood that what ought to be the case cannot be derived from what is the case. As much is true for the alethic modalities. An argument to the conclusion that something must happen requires the requisite sense of obligation to be embedded in its premises. Whereupon the laws of nature are again seen barging dramatically into human affairs, unneeded and so unwanted. On tripping—it happens—I can quite well fall toward the center of the earth without their help. The laws of nature do not result in, produce, conduct, convey, control, or otherwise have a hand in human action—or in anything else. They are, those laws, not events; they do not enter into causal relationships; they are no more bound to the wheel of time than the numbers and mathematical functions that they contain. They are, after all, propositions, expressed in ink or hammered into computer code or carried on a woman's warm breath, a part of the larger linguistic tapestry by which the world is described and explained.

They do what they do. The world is what it is. Nothing is governing anything.

## Freedom of the Will

To the extent that he is eager to say that he does not believe in free will, Brian Greene does not believe in free will.[15] He is hardly alone. It has become fashionable to say as much. The ensuing renunciation is very often considered a manly kind of purgative, like a round of castor oil or a high colonic. Having been persuaded that he has no say in any matter, Greene feels himself curiously obliged to keep saying so, and is pleased to trace the paternity of the least of his remarks backward to the throat of the Big Bang. If free will does not exist, then arguments about free will are, of course, rather like arguments among billiard balls about which pocket they might join. They cannot help themselves, so the arguments are pointless; but neither can they decide to stop arguing, and so the arguments are endless.

Greene is persuaded that he has seen through all this; he is taken in by no disguise. "Here is the modern version of the argument," he writes, "that knocks free will back on its heels." He continues:

> Your experiences and mine seem to confirm that we influence the unfolding of reality through actions that reflect our freely willed thoughts, desires, and decisions. Yet, maintaining our physicalist stance, you and I are nothing but constellations of particles whose behavior is fully governed by physical law. Our choices are the result of our particles coursing one way or another through our bodies. Our actions are the result of our particles moving this way or that through our bodies. And all particle motion—whether in a brain, a body or a baseball—is controlled by physics and so fully dictated by mathematical decree. The equations determine the state of our particles today based on their state yesterday, with no opportunity for any of us to end-run the mathematics and freely shape, or mold, or change the lawful unfolding. Indeed, following this chain ever further back, the big bang is the ultimate source of all particles, and their behavior over cosmic history has been dictated by the nonnegotiable and insensate laws of physics, which determine the structure and function of everything that exists.... We are no more than playthings knocked to and fro by the dispassionate rules of the cosmos.[16]

There is a note of distress in this long paragraph that it would be uncharitable to dismiss. No one wishes to be knocked back, or knocked about, or knocked out, or, God forbid, knocked off.[17] It is a paragraph that conveys a kind of fretfulness, as well, for it offers the coldest and meanest kisses at famine prices—those set at the big bang. If the distress is earnest, the prices are outrageous, the more so since Greene has gone wrong in calculating them. The great physical theories do not say a word about freedom of the will. Newton's laws of motion appeal to time, distance, and mass. This is just enough to make a real world rise. Human life is annotated in other terms and by other words: agency and intention, desire and belief, love and hopeless longing—*sunt lacrimae rerum et mentem mortalia tangent*. Common sense suggests that men act freely when they can, and when they cannot, they do not. This is as far as common sense can go and, for the moment, we have only common sense to go on.

Conflict arises when the discussion turns to theories about theories, theology self-applied, as in the Talmud. If theories in physics are deterministic, does it follow that human freedom does not exist? Greene's affirmative conclusion is full of the drama of his depression. It is doleful: "We are no more than playthings knocked to and fro by the dispassionate rules of the cosmos." The sentiment that these words express is hardly new; it owes nothing to modern science. *As flies to wanton boys, are we to the gods. They kill us for their sport.*[18] No one convinced of the existence of freedom of the will is in the least disposed to doubt that the cosmos very often has it in for us. Not me, at any rate. Still, if the cosmos is often at our throat, it is sometimes at our side. Rough justice, I suppose, but Greene needs more than justice, rough or otherwise, to convey his argument, his confidence in its conclusions notwithstanding.

Schrödinger's wave equation is a well-posed linear partial differential equation, and, as such, completely deterministic, one state rolling after the other to the end of time. When applied to the quantum world in which various particles brag and bounce, Schrödinger's wave equation acquires a novel incarnation. Having once been deterministic, it now becomes probabilistic, the Born rule specifying the probability distribution of various experiments. Those probabilities go all the way down. There is no getting rid of them. Nor do they come as much of a surprise. We live amidst the inconvenience and distraction of the real world, and what Greene has to say has long been said before: *I returned, and saw under the sun, that the race is not to the swift, nor the battle to the strong, neither yet bread to the wise, nor yet riches to men of understanding, nor yet favor to men of skill; but time and chance happeneth to them all.*[19]

What *is* a surprise is the strong conclusion that Greene draws from all this. "Much like Newton," he argues, "Schrödinger leaves no room for free will."[20] No room? Really? Quantum mechanics leaves room enough hospitably to accommodate any number of gabbling philosophers. The place is a polemicist's paradise. Let me see. An agent is free if he could have done otherwise. Thus Greene. He could have done otherwise, that agent, only if things might have been otherwise. Thus me. The laws of probability uphold the possibility that things might have been otherwise. It is of their essence. A fair coin lands on its face only if it might have landed on its tails. Thus logic. This is not yet a defense of free will. If a man might have done otherwise, it hardly follows that he could have done otherwise. Still, it is a step in the right direction. *Il n'y a que le premier pas qui coûte.*

It hardly helps Greene's case, or the cause at large, that freedom of the will has forever defied analysis, even in terms that have nothing to do with physics. Never mind what might have been. Is an agent acting freely if and only if he *could* have done otherwise?[21] On making a nuisance of himself by jiggling his feet in a concert hall, a physicist rarely responds to his irritated neighbors—me, for sure—by insisting that he could not help himself. Such an insistence would suggest that he would have stopped jiggling his feet if he could have stopped jiggling them. It is a charitable suggestion. But if he could not do as much, it follows that he would not have done as much, and this does not appear obviously true. He would have stopped jiggling those fat feet of his had I had anything to do with it.

## Consciousness

In Greene's hands, consciousness is, in short order, promoted to the problem of consciousness, and then to the hard problem of consciousness. The problem is hard, Greene believes, because although "I have direct access to my own inner world, I am… at a loss to understand how that world emerges from the motion and interaction of my own [elementary] particles."[22]

If Greene is at a loss, David Chalmers is hopelessly flummoxed and has begun companionably contemplating the consciousness of rocks. It is possible that the common English expression *dumb as a rock* requires revision.[23] The idea that I have direct access to my own inner world suggests a distinction between inner and outer worlds.[24] The outer world requires indirect access, the inner world is there for the taking. This is hardly the basis for a refined analysis, for it makes a claim that is true only occasionally and only in part. My access to the outer world is often direct, or, at least, as direct as it could possibly be, as when I notice *your* fist approaching *my* nose; and my access to the inner world is, as often, indirect, complicated, a matter of hints, hunches, and inference as much as anything else. In *Of Human Bondage*, Somerset Maugham writes of his protagonist,

> The truth came to him at last. He was in love with her. It was incredible....
>
> He tried to think when it had first come to him. He did not know. He only remembered that each time he had gone into the shop, after the first two or three times, it had been with a little feeling in the heart that was pain; and he remembered that when she spoke to him he felt curiously breathless. When she left him it was wretchedness, and when she came to him again it was despair.[25]

Philip Carey is at a loss in accounting for the history of his own emotions, the odious Mildred appearing in his memory, and in his life, under two quite different systems of interpretation. There is in this nothing unusual. We know what Maugham meant. A man may recognize his emotions only long after the fact, discovering a cold clear core of contempt beneath his insincerely held romantic attachment; he may adjust his judgments and not know whether he is giving up his belief that only the good die young or whether what he really believed was that only the young die good; he may struggle to determine what he is feeling and discover that it is all a hopeless tangle, or he may begin in a hopeless tangle and discover under the imperative of action that what he felt, he had felt all along. This is the way things are.

A word appearing is an error in prospect: our access is always indirect. Beyond the words that we use, what other form of access do we have? The gross topological metaphor by which Greene has divided the world into an interior ball of immediate experience surrounded by a remote, barely glimpsed exterior world of public incidents, events, other people, and ever-receding landscapes is absurd.

Greene is a sophisticated mathematician, and although there is no assurance that mathematical ideas will prove of relevance in analysis, common sense suggests that they could not hurt. Certain emotions very often seem as if they follow a linear and even a one-dimensional order: nothing, nothing, sentiment, *hey*! passion, possession, disappointment, discontent, depression, nothing, nothing, *zu Asche zu Staube*. Other emotions are otherwise. Quite before assigning to the laws of physics dominance over everything, it would be helpful to have a more refined analysis of the alien territories it is claiming. No one is satisfied with the vocabulary on hand; but no one has a better vocabulary either.

Greene considers the philosopher's back and forth with engaging respect, but his heart is not in it. The details he is prepared to leave to others. It is the grand metaphysical question that occupies his attention. How do the sentiments, emotions, inclinations, wishes, and desires emerge from a seething mass of elementary particles? Why stop there? How does anything emerge from something? We do not yet understand how gravity emerges from the elementary particles. It seems rather exigent to demand that those particles somehow get together to cobble together something covering the taste of tea. Emergence makes professional sense only against the background of a theory, so that the requisite inference is one that coordinates a theory of the elementary particles and a theory of consciousness. There is no such theory and so no intellectually respectable bridge between mindless particles and mindful experience. If this is so, it is difficult to see why the hard problem of consciousness is hard; or why, for that matter, it is a problem. A problem incapable of solution is not a problem; and if this suggests that the laws of physics might well be incomplete, why should this come as a surprise? They are already incomplete. The spectral gap problem is undecidable.[26] Greene is at a loss when it comes to understanding how any of this is possible. And so are we all.

## Under the Astrologer's Tent

As the ruler of the soul, Ptolemy wrote in the *Tetrabiblos*, Saturn has the power to make men sordid, petty, mean-spirited, indifferent, mean-minded, malignant, cowardly, diffident, evil-speaking, solitary, fearful, shameless, superstitious, fond of toil, unfeeling, devisors of plots against their friends, gloomy, taking no care of their body. We know the type. Some men are just rotten.[27] Brian Greene is under the astrologer's tent. Ptolemy's heavy arm is draped in friendship over his own frail shoulder. From Democritus to Steven Weinberg, a great many physicists and philosophers find themselves there. Einstein, too. They disagree about the particulars of planets and particles, but not about the chief thing, the idea that something must account for everything.

The astrologers know perfectly well that everything encompasses a lot. They know, too, that as astrologers, they are obliged to trace a connection in nature between something and everything, a form of force or influence, a tangible, if tentative, line. They have not been reticent. They are full of ideas. Ptolemy appealed to a radiation of sorts proceeding from Saturn, and Al-Kindi, writing in the rosy springtime of Islamic philosophy, to stellar rays. Troubled by action at a distance, Albert the Great knew better. A lighthouse in Italy, he remarked, *cannot* influence a lighthouse in England. Brian Greene has made his appeal; he has pitched his case. The elementary particles are fundamental. They are the something that explain everything. If his own discipline of theoretical physics happens to have a prominence denied Assyriology, he may be forgiven a shiver of satisfaction. But what is the connection between something and everything? It is an obvious question. Beyond saying that the laws of physics govern the elementary particles and everything made of the elementary particles, Greene has nothing more to say and says nothing more about it. No wonder. Greene was born under the sign of Aquarius. We are like that, we Aquarians, taciturn.

It requires a certain coldness to pick all this apart. Nothing in nature is fundamental to everything. If one is doing particle physics, the elementary particles are more fundamental than contracts; in drafting a contract, it is the other way around. Quantum fields are the cynosure of quantum field theory; they do not count in physiology. Philosophers hoped that it would be otherwise and that some set of objects would flaunt proudly their fundamental character, but what is fundamental is inevitably a relative judgment, partial, incomplete, and always changing. If a set of objects is fundamental, it cannot explain everything; if it explains everything, it cannot explain anything. This is not a paradox. It is the way things are.

In the world as it is, there is no relationship in nature answering to causality. Thus al-Ghazali and thus David Hume. Neither is there a relationship in nature between the elementary particles and the world or worlds in which they are embedded.[28] These metaphysical imputations have all dwindled and disappeared. What remains is a relationship between *theories* and their *models*, and not between elementary particles and things—bags or otherwise. Inference is the source of influence, and beyond inference, there is nothing. Is there a relationship in nature that corresponds to the inferential relationship between theories? Not obviously. Hardly ever. To have discovered this is among the great achievements of Western science, but it has come at a price, a profound withdrawal from the world. Under the astrologer's tent, a sense of gregarious gaiety yet prevails. Ptolemy's arm is not yet heavy on Brian Greene's shoulders; but deep down the astrologers know that the gaiety will not last. The world in which they could cast spells and conjure with action at a distance has disappeared. What remains is the logician's cold light, cold comfort, I suppose, but better than no comfort at all:

But this rough magic
I here abjure, and, when I have required

Some heavenly music, which even now I do,
To work mine end upon their senses that
This airy charm is for, I'll break my staff,
Bury it certain fathoms in the earth,
And deeper than did ever plummet sound
I'll drown my book.[29]

# 18. Blind Ambition

I BELIEVE. I WANT. I DO. WHAT COULD BE SIMPLER? INTELLIGENCE IS the overflow of the mind in action. In dreaming or desiring, on the other hand, I occupy a world bounded entirely by memory, meaning, and belief: I need *do* nothing. That overflow is entirely internal. In either case, our intelligence is *directed* toward specific objects or states of affairs. I believe—what? That *Clearasil Starves Pimples* or that *Pepsi Is the Choice of a New Generation*; I desire—what? That the young Sophia Loren might step smoldering from the television set for perhaps an hour or that I might win a MacArthur Fellowship (the academic equivalent of the Irish Sweepstakes). What *I* believe (or desire) and what *is* believed (or desired) are connected by something very much like an intentional arrow, a kind of miraculous metaphysical instrument. The relationship between my thoughts and their objects is thus strange from the first. But this relationship between what I think and what I think *about* is duplicated in language itself: like the thoughts that they express, the sentences of a natural language transcend themselves in meaning.

In seeing things from a first-person stance, with the entire world revolving around my own ego—a kind of Ptolemaic system in psychology—I direct the arrow of intentionality from the inside out, infusing the objects and properties of the external world with all of the significance that they ever possess. I assume, of course, that others do as much. Read forward, the arrow of intentionality goes from what I feel to what I do; read backward, from what is done to what is felt. The sense that we are all in this together arises only as the result of a supremely imaginative kind of back-pedaling; the interpenetration of two human souls, when it occurs, is wordless.

There is more. Each of us acts in the world as both subject and object: we do, and things are done to us. In moving away from the lunatic solipsism in which my ego exists in the absence of all others, I endow those human beings in my own perceptual ken with more or less the same cognitive states that I myself enjoy. This is the basis for a sense of sympathy. The endowment itself, I presume, may be reversed, as when I myself figure in someone else's awareness as an imaginatively constructed subject of experience. But here is a queer, artful point. The inferences that I make about others, others make about me. My inferences about others I cannot verify, but their inferences about *me* represent something like the backward wash of a familiar wave. A subject acting simultaneously as a psychological object enjoys a unique Archimedean perspective on the system of inferences by which mental life in the large is constructed.

This confluence of circumstance suggested to the American philosopher John Searle a very deft argumentative maneuver, something akin, really, to a movement in *judo*. His arguments were prompted by work undertaken at Yale by the psychologist Roger Schank. Like many other American theorists, Schank has approached the problem of artificial intelligence with a kind of bluff, no-nonsense sense that getting a machine to understand something is a matter of attending to the details in a patient, straightforward way. In a photograph at the back of his book, *The Cognitive Computer*, he stands with his arms folded over his ample belly, scowling directly into the camera, an expression of earnest ferocity on his face, as if to suggest that by the time *he* got through with them, those computers of his would either shape up or ship out. His aim, as he explains things, is to teach the digital computer to comprehend simple stories of the sort that might be told to children.

The exercise is set out without irony. The education of the digital computer in this regard commences with what Schank calls a *script*—a kind of running, rambling background account in which the saliencies of various stories are set out and explained. With the scripts in hand, the computers are prepared to make sense of what they read. They are then interrogated with a fine eye directed toward telling whether they have understood what they have absorbed. In fact, Schank's machines *do* get quite a bit right; the record of their conversation is admirable, and the unbiased reader often has the feeling that just possibly he is reading something strange and remarkable.

It is against this conclusion that Searle has set his face. It is a simple fact, Searle begins, that he is utterly ignorant of the Chinese language. Suppose that he were to be locked in a room with a large sample of Chinese script—the samples, say, arranged on cardboard sheets. Now imagine that Searle were to be given "a second batch of Chinese script together with a set of rules for correlating the first batch with the second." The rules are in English. A third collection of scripts is presented Searle. And another set of rules. This makes for three separate sets of Chinese symbols and two sets of English rules.

From Searle's point of view, the material he confronts is an incomprehensible jumble. From the outside, where sense is made of all this, those Chinese symbols have a specific meaning. The first corresponds to a general script—the sort of thing that a computer would need in Schank's setup to make sense of a story. The second is actually a story in Chinese. The third represents a list of Chinese questions. From time to time, those questions are presented to Searle with a nudge and a wink and a tacit request that he say something. In answering, Searle consults his set of rules. The two sets enable Searle to match the questions to the story by means of the background script. In this respect, Searle remarks, he is precisely in the position of the digital computer.

But (a very excited, explosive *but!*) under such circumstances would there be any inclination to say that a subject so situated understands the meaning of the symbols he is manipulating? An *observer* might come to this conclusion. Put a question in Chinese to this character, after all, and he answers in Chinese. Yet this is not at all how Searle himself sees things. Whatever he may be able to *say* in Chinese, he remains confident that he *understands* nothing of what he has said and is prepared to champion his ignorance defiantly. Some great notable aspect of what it means to understand a language has simply been overlooked.

For the most part, computer scientists have tended to ignore Searle's argument and the point of view that it represents. It had long been known in science that you cannot beat something (a research grant) with nothing (a destructive argument), and what Searle had to offer them was nothing at all. Analytic philosophers responded promptly to Searle. The results are confusing. A great many superbly confident rebuttals appear to contradict one another. As for myself? When pressed on the point, I tend to run my hands through my hair or tug mournfully at my ears, gestures I am convinced that suggest that I have something tack-sharp to say were I willing only to say it.

# 19. KOLMOGOROV COMPLEXITY

THE AIM OF SCIENCE, RENÉ THOM ONCE REMARKED AS WE SIPPED espresso in a café near the Opéra, is to reduce the arbitrariness of description. I nodded my head and fixed my face in an expression that I thought conveyed a sense of alert but sophisticated appreciation. On the view of things that I had been taught at Princeton, the aim of science, insofar as science has any aims whatsoever, is a matter either of explanation or prediction. Going further, explanation involves seeing in the particular (this swan is white) intimations of the general (all swans are white) in such a way that the particular, when properly described, follows deductively from the general. In moving upward (past the swans, at any rate), the scientist ascends toward the laws of nature.

I thought to ask Thom what he meant, but by the time I had posed my question in a way that suggested I knew the answer, Thom was industriously applying himself to an apple tart.

Science is pursued, I think, for many reasons, not the least of which is to fill up the time. In this regard, it is always successful. Insofar as science is purely an abstract activity, like mathematics, chess, or nuclear strategy, it is undertaken chiefly for the acquisition of that magical moment in which things that formerly stood distinct and separate fall together in a limpid whole. Such is intellectual bliss—paler by far than physical bliss, but nothing to sneeze at either.

Blisswise, certain concepts form a tight circle in the sense that each may be used to define and justify the other. *Complexity, information, randomness, order,* and *pattern* (or *form*) are connected like the members of a family of cheeses: Gruyère, Brie, Port Salut, Camembert, but not Velveeta. This suggests that they may all flourish, or fall, together. Shady characters of all sorts—semioticians, anthropologists, linguists, sociologists, communication theorists—are especially partial to concepts of just this kind, perhaps because of the way that their names fill the mouth when uttered. This is no reason to reject these concepts out of hand, but no cause for congratulation either.

The technician, or the astrologer, no less than the rest of us, is pattern intoxicated. Reading the charts or the stars, he sees the subtle seams by which nature is constructed—the pattern at the bottom (or top) of things. A pattern is peculiar in that knowing part (moving in) one is likely to guess correctly at the rest (moving out). The patterns scouted by the stock market technician are especially plain: if they are there at all, they are there on the surface of things. On the other hand, consider the numbers 1, 4, 1, 5, 9, 2, 6, 5, 3, .... There is not much by way of pattern here; still less when the sequence is extended: 5, 8, 9, 7, 9, 3,.... A cursory examination might suggest that these numbers are quite without significance. Not so. They represent the decimal expansion of *pi,* to use an example that, like Mexican food, keeps coming up. Here the pattern is a matter of the way in which the sequence is *generated* and lies hidden from the surface.

There is pattern, then, and generative pattern. Suppose the world contracted to a pair of symbols: 0 and 1, say. A binary sequence is a system of such symbols in a distinct order—0, 1, 1, 0, 1, 1, for example—and of a specified length—six in the present case. Six binary symbols may be arranged in 64 separate sequences. In the general case, a sequence of length *n* (there are *n* symbols) may be recast in $2^n$ separate ways as $2^n$ separate sequences. Sixty-four is just $2^6$, where $2^6$ is 2 multiplied by itself 6 times.

Imagine now that binary sequences are being produced at random— by the action of a roulette wheel, for example. Of the two sequences

1) 0, 0, 0, 0, 0, 0,

2) 1, 0, 0, 0, 1, 1,

the first seems distinctly less likely than the second: a man idly flipping coins does not expect to come up with a run of six heads. Yet in point of probability, the two sequences are reckoned alike. There are sixty-four possible sequences in all. Each has a 1-in-64 chance of occurring. The most natural probability distribution over the space of n-place binary strings assigns to each string the same probability—$2^{-n}$. It goes against the grain, mine, at any rate, to accept this conclusion, especially when $n$ is large; but nothing *in the sequences themselves* indicates obviously that one is less (or more) likely to occur than any other.

Sometime in the 1960s, Russian mathematician Andrey Kolmogorov thought to argue that the degree to which a given binary string is random might be measured by the answer to a simple question: to what extent can the string be re-described? Kolmogorov thought of the possible re-descriptions of a given string as instructions to a fixed computer. Now if $S$ is a binary string its length is measured in bits. An $n$-place binary string is $n$ bits long. The most obvious re-description of $S$ is $S$ itself—the sort of thing I might send you to make sure that you get what I mean. In the case of sequences such as 2, nothing less will do. 1, on the other hand, may be expressed by a single terse command: Print 0 six times. A simpler description of binary string is thus a *shorter* description of the string. Sequences that *cannot* be generated by shorter sequences, Kolmogorov argued, are *complex* or *random*. This is a definition. But random sequences are precisely those that are rich in information. The definition thus ties together four concepts loitering casually at the margins of this discussion: randomness, compactness, complexity, and information. Playing an unusually inconspicuous role is the notion of probability.

Kolmogorov first spoke on this subject in a brief note published in 1967. His work was duplicated by the American mathematician Gregory Chaitin, who experienced a flash of intellectual lightning while an undergraduate at the City University of New York, sitting among students baffled by long division. The subject is known now as *algorithmic information theory*. Those algorithms are a reminder that Kolmogorov thought of descriptions in terms of inputs to a fixed computer.

Quite surprisingly, the problem of decisively determining whether a given string is random turns out to be unsolvable. If a shorter description of the string may actually be produced, well and good. If not, all bets are off. A shorter description may exist; then again, it may not. There is no demonstrative telling. The *decision problem* for complexity is recursively unsolvable. Like truth, randomness is a property that remains ineluctably resistant to recursive specification.

Kolmogorov's elegant and simple idea—a little jewel, a diamonoid—achieves its startling effects by means of an especially simple series of inferences. If all else fails, a binary sequence of length $n$ may be re-described by a binary sequence of just the same length. There are $2^n$ such sequences, and $2^n - 2$ sequences shorter than this. But on any reasonable interpretation of complexity, sequences within a fixed integer $k$ of $n$ itself must be reckoned random or complex if the $n$ - place sequences are themselves reckoned random or complex. It follows that only $2^{n-k} - 2$ sequences are less complex than $n - k$. If $k = 10$, the ratio of $2^{n-10}$ to $2^n$ is precisely 1 in 1,024; the ratio of simple to complex sequences is thus on the order of 1 in 1,000. This means that of 1,000 sequences of length $n$, only *one* can be compressed into a program more than 10 bits shorter than itself. The number of purely random sequences grows exponentially with $n$, of course, and this implies that randomness and complexity are the norm in the general scheme of things. But if *most* sequences are random, the appearance of 1 *should* prompt a natural sense of surprise; sequences like 2 are what one expects and what one generally gets.

This line of argument, of course, resolves one problem only by embedding it within another—resolution by delayed dismay.

Metaphysics, the reader may have guessed, is not entirely in my line. I am tempted to cross over just this once. On Kolmogorov's definition of complexity, sequences are simple if they can be briefly described. Science is itself a matter of data, success in science a mystery of abbreviation. A law of nature is (among other things) data made compact: $F = ma$, said once and for all, the whole of an observed or observable world expressed in just four symbols. The fact that science is only partially successful suggests that only parts of our experience are regular. As for the rest, there one confronts a resolute kind of amorphousness—something in-eradicably resistant to scientific specification. It is the noble assumption of our own scientific culture that sooner or later everything might be explained: AIDS and the problems of astrophysics, the life cycle of the snail and the origins of the universe, the coming to be and the passing away. It is not possible to contemplate this aspiration with anything but *Attaboy* on one's lips. Yet it is possible, too, that vast sections of our experience might be so very rich in information that they stay forever outside the scope of theory and remain simply what they are: unique, ineffable, insubsumable, irreducible.

# V. Deep Dive

# 20. A Review Of Michael Ruse's *The Philosophy of Biology*

THE PHILOSOPHY OF BIOLOGY BY MICHAEL RUSE

HUTCHINSON, 231 PP.

THE NATURAL THOUGHT THAT THE PHILOSOPHY OF BIOLOGY COM-prises a kind of intellectual Lapland owes much to the conviction that biology itself is somehow a derivative science, an analogue perhaps to civil engineering, and in any case devoid of principles fundamental as those that mark the physical or chemical sciences. This is a position that J. J. C. Smart takes to be congenial, arguing for it briskly in *Philosophy and Scientific Realism*; it seems much the view of those molecular biologists who like James Watson look to the biochemical properties of the cellular macromolecules for a complete accounting of the biological properties of the gene. Professor Ruse sets *The Philosophy of Biology* against this line: the "logical empiricist" account of such theories as physics or chemistry afford, he advises, applies in large measure to biology as well.

But Ruse takes a resolutely unmolecular attitude toward modern biology: molecular genetics is mentioned at Chapter 10; points between this and the Introduction are taken with examples drawn from Mende-lian genetics, population genetics, and the neo-Darwinian theory of evo-lution. This gives the book an off-center tilt and has an evil effect on the chief arguments of the early chapters. Thus the skeptic will respond to

Ruse's assurances that the Mendelian gene is an entity of splendid theoretical merits—as obscure in its own way as the electromagnetic field, the valences of carbon, or the now mythical ether—by observing that insofar as the Mendelian gene exists at all it exists in a purely biochemical capacity, a segment of those ropey nucleic acids that give form and content to tons and tons of blubbery protoplasm; the snugness with which the logical empiricist account fits biological thought has little to do with its autonomy and reflects instead the simple if uninspiring conclusion that biology is indistinguishable from biochemistry. Now this is not an argument that I find persuasive, but it would wobble me considerably if I had to make the contrary case by appealing to the theoretical integrity of the Mendelian gene.

Chapter 3 takes the sagas of empiricism to the axiomatization of population genetics—*Larousse Axiomatique*. By my lights the reregistration of scientific theories as axiomatic systems in either standard or set-theoretical formulation reflects a barren architectural passion. Ominously enough, Ruse writes that "to say that something is axiomatized is to say that we start with some statements as premises… and from these we derive other statements,"[1] a formulation of transcendental inadequacy yielding, in the record of Ruse's aspirations toward the derivation of the Hardy-Weinberg law from Mendel's law of segregation, evidence of anorexigenic slightness in favor of the thesis that biology is as axiomatizable as any of the sciences—physics, say. What is worse, Ruse is an *Innocent of Riger* (*naïf logique*): the logical details are imprecise, the principles that *entail* the Hardy-Weinberg law in Chapter 3 constitute its chief stock of evidence by Chapter 6.

The Hardy-Weinberg law itself is something of a triviality, made notable only in virtue of the steadfastness with which early geneticists endorsed its denial. The redoubtable Udny Yule, for example, writing in Volume 1 of the proceedings of the Royal Society of Medicine, invoked a genetic version of Gresham's Law against the Mendelians: if brachydactyly is dominant, he insisted, "in the course of time one would expect, in the absence of counteracting factors, to get three brachydactylous persons to one normal." But against this, Hardy argued that, suppose one is considering three genotypes, $AA$, $Aa$, $aa$ occurring in a randomly mating population in the ratios: $u:2v:w$. Set $p = u + v$ and $q = v + w$—this on the assumption that $u + 2v + w = 1$. Clearly the ratios of $p$ to $q$ will reflect the ratios of $A$ to $a$. If maternal and paternal genes are selected independently, the probability that an offspring in the first filial generation will be $A$ is obviously $p^2$: first generation genotypes occur in frequencies of $p^2:2pq:q^2$. Now imagine a fixed distribution of genetic frequencies $p$ and $q$ such that $p + q = 1$ and consider any distribution of genotypical frequencies such that $p = u + v$ and $p = v + w$. Genotypical frequencies in the first filial generation will all occur in the ratios: $u_1 = p^2$, $2v_1 = 2pg$, $w_1 = q^2$. Moreover, at least one distribution will be *stable* in the sense that $u = u_1$, $v = v_1$, $w = w_1$. It will take at *most* one generation to reach a stable distribution; for such populations there is no chance at all that the dominant genes will drive out all the rest.

Ruse says as much in setting things out systematically, but his long and chart-crowded discussion is an exercise in misapplied force, like a screw hammered into wood. He considers a sample population and assigns arbitrary frequencies to heterozygous and homozygous genotypes: $P\ A_1A_1$: $H\ A_1A_2$: $Q\ A_2A_2$; the ratios of $A_1$ and $A_2$ are as $p$ to $q$. On the natural assumption that mating in a population neatly divided along sexual lines is random, the frequency of mating types can be computed as a simple product of genotypical frequencies. There are nine mating possibilities in all and for each it is simple enough, given the nature of the crosses, to reckon the likely frequencies associated with their offspring. So setting things out in tabular form, with mating frequencies along the side and projected progeny across the top, a simple summing of the resulting columns will give the overall frequencies with which each of the three genotypes may be expected to reproduce. On the assumption that $p = P + \frac{1}{2}H$ and $q = Q + \frac{1}{2}H$, $P_1$, $H_1$, and $Q_1$ will stand for each other in the ratios: $p^2:2pq:q^2$.

This is precisely the observation to which the incautious Yule provoked Hardy, but Ruse might have compressed the three pages that his argument takes by beginning with the relationship $p = P + \frac{1}{2}H$ and then following the line I took following Hardy. This would not have brought about a resuscitation of Ruse taken under the aspect of axiomatic: my own argument was inconsequentially informal, with Mendel's first law coming into play as governing assumption and not as a single axiom from which a plumb line to the Hardy-Weinberg law might be dropped.

The discussion also suffers from an infelicity in stress: in his great-hearted enthusiasm Ruse comes to excuse the Hardy-Weinberg law, arguing that like Newton's first law of motion it seems to say simply that if nothing happens "then everything will stay the same."[2] This is hopelessly inventive, but the Hardy-Weinberg law exudes a sense of yawning obviousness just because it is a truism; one looks to it in astonishment if only for evidence that a given population, whatever its genotypical make-up, will achieve reproductive stability within a single generation; this is not a point that Ruse makes effectively, and the reader is left with the impression that biologists have curiously chosen to celebrate a formula that gives unassailable expression to the fact that breeding populations, if large enough, obey the laws of probability.

Still Ruse rightly rejects the view that there is something conceptually indecent about the philosopher who conceives a longstanding and inexpungable passion for an axiomatic biology. It was J. J. C. Smart who fashioned Smart's *animadversions against the axiomatics*: what is distinctly biological, he argued, is generally not lawlike, and what is distinctly lawlike, he added, is generally not biological. Either way the chances for an axiomatic biology are poor. The whole of Smart's objections has come to be known as *Smart's lament* and Ruse rejects it for calling on a concept of lawlikeness of indefensibly ferocious narrowness.

Nor are there irremediable deficiencies in the logical structure of mathematical genetics; the reader who needs a sense of the sophistication of the subject should see the Appendix to Kimura and Ohta's brilliant *Aspects of Population Genetics*. There Kimura and Ohta take as central the Kolmogorov forward and backward equations, and show that the wandering of a gene from one frequency to another satisfies the Kolmogorov forward or Fokker-Planck equation, which they in turn derive from reasonable probabilistic assumptions. I see no intrinsic differences between the intellectual structures of mathematical genetics and solid state physics; on the issue of lawlikeness I take a position of perfect profligacy, counting the Fokker-Planck equation, if it holds, as a law on all fours with the laws of physics and turning a deaf ear on all claims that the only laws worth having are those of the very most fundamental parts of physics.

So far I go with Ruse, but no further; for while he hails biology—population genetics, at least—as comparable to chemistry in virtue of its axiomatic ripeness, I take the position that the habitual invocation of the axiomatic method is a mistake; and while I admire the philosopher taking formal methods to subjects precise with the limits of ordinary mathematical practice, my enthusiasm for his zeal is corrupted by my indifference toward his aims.

With molecular biology it is harder to respond with superb unconcern to calls for the devaluation of biology. The chief results are plainly biochemical; cells obey the laws of chemistry. Chemistry is where the laws are, so axiomatizing the applications would be something like axiomatizing the bottom half of set theory. This is an argument that leaves Ruse unfazed, but then, he fixes the center of biological thought at a point that sees a mingling of population genetics and the synthetic theory of evolution, a view of distinct regional oddity, rather like the claim that Cudworth is a more important philosopher than Locke. But even for things seen correctly with molecular biology exhausting not only the center but *all* that is interesting in biology, the withering away of the intrinsically biological in favor first of chemistry and then of physics bulks large but weighs little. Living creatures are just biochemicals variously arranged, but this gives no hint of the unfathomable complexity that the arrangements may take.

B. L. Hendricks coined the term "natural objects" to signify those entities admitting of explanation in terms of the objects that compose them; some structures—the human or mammalian immune-defense system, for example—may be unnatural in the sense that analysis of their capacities in terms of the molecules that make them up might go beyond our biochemical abilities. E. C. Zeeman, aghast no doubt that biochemists spend the better part of their working lifetime in the analysis of a *single* molecule, turned in relief to theories that stick to the conceptual surface of things. In analyzing cardiac nerve impulses Zeeman proposed a global account of their dynamical properties; his own work is an application of René Thom's catastrophe theory; the underlying aim of the whole business is to stay clear of the biochemistry and work instead with features that can be represented by theories drawn from differential topology or the qualitative theory of differential equations.[3] There *is* something unsatisfying about theories that fail to force themselves down to the level of those constituents that account finally for the way things are; but given the frailty of our intellects and the complexity of the data, there may be natural but not logical limitations on depth.

Chapters 4, 5, and 6 pit the doctrine of natural selection and the neo-Darwinian theory against criticism of markedly limited natural aggressiveness. "Evolution is," Ruse writes, "the result of natural selection working on random mutation"[4]; biologists take the theory as an imperishable glory, so there is something vaguely disagreeable, Ruse believes, about the fact that so many philosophers "think that (its) basic truth is open to question."[5] But critics have argued that the theory is *empty* rather than false and biologists discuss its harmonies by first describing as imperfect the interpretations of their colleagues.

An obvious sticking point to the theory is the concept of *fitness*. If by "the fitter organisms" biologists mean merely those that survive, then the doctrine that selection winnows out those that are fit achieves a kind of languid triviality. Biologists indignantly reject this reading of the theory: Ernst Mayr, for example, asserts that to say "this is the essence of Darwin's reasoning is nonsense"[6]; an attentive reading of Chapter 8 of his own justifiably appreciated text shows that it is an analysis to which Mayr himself is partial. In writing of traits that have no apparent selective advantage, Mayr argues that "the mere fact that such traits have become established makes it highly probable that they are the result of selection, that is, of an unequal success of different genotypes."[7] If this does not mean simply that those traits that survived, survived, then whatever else it might mean is conceptually ellipsoid.

Mathematical geneticists also write as if fitness were one property, reproductive successfulness another. The hypothesis of the theory of evolution, they will say, is just that organisms exhibiting the former enjoy the latter; in the writings of Fisher, Haldane, Wright, and more recently Kimura and Feller, the notion of fitness assumes an air of imperturbable propriety. There is talk of fitness levels, fitness profiles; all manner of calculations are made featuring fitness as a parameter; under certain combinations and permutation of the genetic pool, geneticists even envisage situations in which the fitter stock is washed out of a population, with inferior genetic material spreading like a stain over the alleles due to the action of genetic drift. For all that, the notion of fitness retains something of the shadow. The relationship between fitness in the sense of population genetics and fitness in the ordinary biological sense is much like the relationship between the concept of stimulus in mathematical stimulus sample theory and the ordinary empty notion of stimulus in behavioral psychology.

The trouble has to do with the rather meager possibilities for the independent assessment of fitness. On the accounts of mathematical geneticists, fitness is a property admitting of a measure, but what is actually measured, it turns out, is the frequency with which a given allele reproduces itself: a selective edge comes to nothing more than a reproductive advantage. Feller, for example, considers the case in which two genes $A$ and $A'$ vary in fertility by $\mu$ and $\mu'$. The ratio $\mu/\mu' = 1 - k$, where $k$ is a coefficient of selection, represents the relative fitness of $A$ and $A'$;[8] but plainly whatever it is that accounts for the difference between $\mu$ and $\mu'$ has become an impalpable power, registered only through its effects. The philosopher who wants to know why a species that represents nothing more than a persistent snore throughout the long night of evolution should suddenly (or slowly) develop a novel characteristic will learn only that those characteristics that survive, survive in virtue of their relative fitness; those characteristics that are relatively fit are relatively fit in virtue of the fact that they survived.

There are, it is true, occasions when one can appeal to cases of successful adaption and make plausible the assessment of a particular trait as an especially valuable one. Ruse cites Kettlewell's work on the speckled moth (*Biston betularia*) and then, as a *gedankenexperiment*, imagines a mammalian population afflicted with what seems to be elephantiasis: if mammals with legs swollen to the size of silos were to outstrip the others in reproductive zeal, this would weigh heavily against the neo-Darwinians. There is no refuting a theory that is empty: if the critics are right and neo-Darwinism a fast blur, no sense could be made of such falsifying circumstances as Ruse proposes.

Yet, Ruse comes to miss the more general point: these inextendable instances show up on principles by which biologists could come to measure the fitness of a vast stock of genetical adaptations. Will a morphologic shuffle that sees the pig switch from trotters to wheels mounted on ball bearings effect an enhancement in its fitness? No one knows, although some guesses are possible. But equally, where along the continuum of fitness are we to fix such historical developments as the tuft on the tufted ibex, the human appendix, or the third toe of the by now legendary three-toed sloth? Ruse's own discussion of this point, in response to difficulties voiced by Manser,[9] suffers from what mechanical engineers call "uncontrolled oscillation." Thus midway through page 40 Ruse argues that "one does not have, as critics frequently claim, a straight identification of those which survive and reproduce with those which are fit." So far so good. But then, in summarizing his position, Ruse says that "those which do have the reproductive edge" biologists "call the fitter," which suggests that an organism to be counted fit must simply survive. Yet Ruse also writes—I am still on page 40—"that given enough organisms a sufficiently high proportion of the fitter are better reproducers, so the fitter members as a whole have a reproductive edge." This evokes a purely empirical construal of the claim that the fit survive. How then are we to reconcile this sensible avowal with the remark that follows paraphrastically: "... this claim," Ruse observes, "is indeed analytic, for this is a definition of what is meant by 'fitter'"?

The theory of evolution brings to light an almost zoological clutch of critics. There are philosophers holding to Lamarckian doctrines, to the saltation theory, to any number of bizarre positions; there is even Professor Himmelfarb taken together with the logic of possibilities. But no mention is made of theories of non-Darwinian evolution or criticism of the sort articulated by Eden, Schützenberger, Gavadan, and Berlinski.[10]

Towards the material he does discuss, Ruse is rather undemanding, with nothing like a feel for what seem to me to be obvious infirmities in the *qualitative* structure of the theory of evolution. Bernhard Rensch, for example, observed that "animals the enemies of which find their prey by the eyes, develop protecting color or shapes."[11] Rensch dubbed his observation *Rensch's Law*; and it has come to be called that in the literature. There are other comparable examples. But Ruse never asks why organisms with different genetical structures should come to essentially the same adaptive conclusions. Are there unique strategies to evolution—configurations that in the nature of things exclude all the others, or options so imaginatively marked that the capable animal could not fail to hit on them as the perfect embodiment of its environmental needs? If so, how does a purely random search turn them up? And if not, why along the margins of the northern tundra is there an obdurate preference for the dubiously effective stripe, the melding blur of fur that from a distance serves to merge the snow rabbit with the snow? Something rhetorical about these questions would no doubt provoke biologists to a blizzard of informed and able responses. But what I am looking for are the *general* and *qualitative* features of the theory of evolution that will make it clear why a given species developed the particular and complicated forms of life that it did; the usual very rapid utterance in exact sequence of the words *random mutation* and *natural selection* strikes me as a gluey conversion of the evidence into an explanation. This, I gather, is how Marcel Schützenberger sees things: the unprincipled virtuosity with which biologists explain bioluminescence in the fireflies or transformational grammars in the mammals suggests that the theory they call on is a paradigm "of what it means to be… nonfalsifiable."[12]

Chapters 7 and 8 are given over to a discussion of taxonomy, a subject that compares unfavorably with Schenker Analysis or the study of Sumerian. Consider, for example, the *monotypic* taxon, which gives rise to Gregg's paradox; the best known case, Ruse reminds us, "involves the aardvarks, whose order, Tubulidentata, includes only one species, so that the order Tubulidentata, family Orycteropodidae and genus *Orycteropus* are all monotypic, containing just the same members as the species *Orycteropus afer*."[13]

Set-theoretically, species and order in the above case are the same— they have the same members. But according to Linnaeus, species and order *cannot* be the same. One has here an irresistible force and an immovable object; the ensuing friction has even brought some taxonomists to the casting off of set theory. I myself would opt for the effective but aesthetically repulsive strategy of putting individuals on one level of the taxonomical hierarchy but unit sets of individuals above them. Then instead of those endlessly iterated aardvarks, one would have the aardvarks first, followed by unit sets of aardvarks, followed in turn by unit sets of unit sets of aardvarks, and so on up to the transcendental aardvark, which I picture as something rather like an inaccessible Cardinal. Ruse rejects related stratagems on page 152 and calls for something more radical: "The axiom of extensionality must be dropped and it must be allowed that taxa of different rank… can have the same members. One can do this by permitting taxa names to have *intensional* definitions, rather than, as set theory requires, extensional definitions."[14] I am not sure what I should count as an intensional definition, but plunging on, I take Ruse as calling for the construal of taxa as *intensional objects*, natural kinds, perhaps, or properties. Given this there is no need to go further and jettison the axiom of extensionality, which says only that sets are identical if they share the same members; and if Gregg's Paradox goes via the observation that two distinct properties may be exemplified by exactly the same individuals, now difficulties arise when we consider how many and which taxa should be instituted in recognition of the fact that many properties coincide. This is not a paradox precisely, and I have not seen it discussed in the taxonomical literature: I propose forthwith to introduce it under the name *Sggerg's Imbroglio*, where "Sggerg" is just "Gregg's" spelled backward to suggest the hidden symmetries between the problems as they are variously posed.

This suggests Ruse wrongly read and, indeed, cutting across the arguments that drift toward the Imbroglio is his assertion that "taxa names can be defined by specifying a number of properties required for taxon membership, rather than by mere enumeration of the members,"[15] which implies that the difference between the taxa is all in their definitions. Thus in the paragraph celebrating the *dissolution* of Gregg's Paradox, Ruse writes that "if we now allow that taxa names may be intensionally defined... although a monotypic taxon will have no more members than its sole included taxon, the two taxa will be kept separate by the fact that the taxa will have different requirements for membership."[16] But this is to confuse the taxa with their names: what will be kept distinct, on this hopelessly implausible construction, are not taxa but definitions; so long as the aardvarks are collected in taxa that as *a matter of fact* are monotypic, species, order, family, and genus will achieve a perfect and irrefragable set-theoretic coincidence.

Chapter 8 carries the discussion of Sokal and Sneath's *Principles of Numerical Taxonomy*, a work that was accurately appraised by Ross as "an excursion into futility."[17] Chapter 9 is an account of functionalism and teleology, perennial favorites in the philosophy of biology, and in Chapter 10, matters are for the first time considered from the molecular point of view. The exposition of "the central dogma" is clear enough, but the discussion sticks pretty much to the question whether Mendelian may be reduced to molecular genetics. A Postscript records Ruse's conviction that "just as at one end biology is merging with the physical sciences, so at the other end biology will merge with the social sciences." This is not a prediction of relieved bleakness; for while the biologist may find himself pre-empted by gibbering sociologists, stodgy biochemists, or prancing behavioral psychologists, given over to unspeakable experiments and inedible prose, Ruse foresees in the Yin and Yang of things "an increasingly important role" for the philosopher, and in these hard times that at least is cause for cheer.

# 21. The Director's Cut

*"The creation of numbers was the creation of things."*
—Thierry of Chartres[1]

It is a useful phrase—*the director's cut*. It is useful because it sets the scene: the masterful director; *his* view; *his* vision. Mathematics has always been rich in its directors of note. In the third century BCE, Euclid subordinated two-dimensional space to five axioms. Twenty-three hundred years later, Giuseppe Peano brought the natural numbers under axiomatic control. The natural numbers are the oldest objects in our intellectual experience. Without them, we would be lost. Peano arithmetic comprises Peano's axioms together with their logical consequences. It is a theory as rich as any in the sciences.[2] The hoarse, excitable Peano is behind the lens. Peano arithmetic, $P_A$, is what he sees; the natural numbers, $\mathbb{N}$, are what they are.

There it is: $<P_A, \mathbb{N}>$, the director's cut.

## Formal Arithmetic

Kurt Gödel published his remarkable incompleteness theorems in 1931.[3] With the exception of John von Neumann, mathematicians found Gödel's work very difficult.[4] It remains difficult, uncanny in its power.[5] "The formulae of a formal system," Gödel wrote, "are, looked at from outside [*äusserlich betrachtet*], finite series of basic signs."[6] The basic signs are what they seem, but who is looking at them from the outside? It can hardly be Peano. The symbols in Peano arithmetic are transparent: Peano is looking *through* them. It is what we all do. A contrived withdrawal is required before symbols can be seen as signs.

Peano arithmetic is expressed in the language of mathematics and in ordinary English.[7] For all natural numbers, $m = n$ if and only if $S(m) = S(n)$. Two numbers are equal if and only if their successors are equal. The exuberant mathematician is handling things with an easy familiarity. And why not? He commands the scene: Peano's axioms, Peano arithmetic, and, beyond them, the natural numbers. The logician stands apart. His is an orthopedic practice. He is concerned to see the articulated skeleton beneath Peano arithmetic, its formal system $F_{PA}$.[8] It is the logician who is looking at things from the outside.

Like certain objects in the physical world, $F_{PA}$ is atomic in nature. Its atoms are its basic signs; its molecules, their combination. Logical constants are first: '∼' (not), '∨' (or), '∀' (all), '0' (zero), '$S$ (the successor of), and, in ')' and '(', right and left parentheses. Although sparse, these signs cover the contingencies, ordinary numerals vanishing in favor of a stuttering '$S$', so that '$S\ S\ S\ S\ (0)$' in $F_{PA}$ does duty for '4' in English. Individual variables are next: '$x$,' '$y$,' '$z$,' ... , and the like. With arithmetical functions, predicates, and relations, it is more of the same.[9] Whatever the signs, they must be combined, and this is a procedure superintended by specific rules. Elementary formulas are given explicitly. The well-formed formulas are then defined as the smallest set containing all of the elementary formulas and their negations, disjunctions, and universal quantifications.

If Peano arithmetic is an arithmetic theory, it is also an axiomatic theory. Theorems follow from the axioms in a gross cascade. It is the logician who must specify the rules of inference of a formal system, the flow of its deductions. An appeal to meaning is not allowed. A formal system is a system of signs. It might as well be a system of shapes. Some inexorable sense of rightness nevertheless prevails. Given

      Mierīgie ūdēņi ir tie dziļākie ⊃ Mierīgie ūdēņi ir tie dziļākie

and

        Mierīgie ūdēņi ir tie dziļākie,

it follows that

        Mierīgie ūdēņi ir tie dziļākie.

If the underlying language is inscrutable,[10] this inference is crystal clear, contingent only on the antecedent identification of '⊃' as a logical particle.[11] The rules of inference governing a formal system are procedural; but as rules of procedure, they correspond, or express, a natural motion of the human mind, the power to move from one proposition to another.

Peano arithmetic is constructed from the Peano axioms,[12] but in the context of formal arithmetic, those axioms lose some of their incidental glamor and appear as axioms in virtue of being called axioms. They have been set apart by being set apart. The inference that has just led to *Mierīgie ūdēņi ir tie dziļākie* is modus ponens, and modus ponens is one of the two rules of inference that figure in Gödel's argument.

The second expresses the common understanding that what holds for anything holds for something. In these cases, to go inferentially from one formula to another is to go somewhere at once. There are no intermediate stops. The provable formulas comprise the smallest set containing the axioms of Peano arithmetic and closed under the release of immediate inference.

In $F_{PA}$, a new mathematical object has come into being. Physical objects are described by the laws of physics. Their nature is disclosed. A formal system is determined by its formation rules, its rules of inference, and by the screening off that screens off its axioms. Its nature is induced. It is induced subject to a double standard of effectiveness. The well-formed formulas, the axioms, and the rules of inference make for three sets of formal objects. Standards for membership are strict. It must be possible to decide whether a given formula is well-formed *and* whether it is not. It must be possible to do both in a finite series of steps. And it must be possible to do as much for the axioms and the rules of inference.[13]

"Whatever withdraws us from the power of our senses," Samuel Johnson observed, "whatever makes the past, the distant, or the future predominate over the present, advances us in the dignity of thinking beings."[14]

## The Recursive Scaffold

Peano arithmetic is a theory about the natural numbers, and these have remained unattended and unvoiced. They now come into their own. Gödel used the natural numbers to identify the formulas of a formal system, the identification so close that Gödel felt justified in referring to the basic signs of a formal system *as* natural numbers.[15] To every basic sign, and then to every formula, and then to every sequence of formulas, and then to every series of such sequences, Gödel assigned a unique number. Basic signs are mapped to basic numbers: '0' $\leftrightarrow$ 1, 'S' $\leftrightarrow$ 3, '~' $\leftrightarrow$ 5, '$\lor$' $\leftrightarrow$ 7, '$\forall$' $\leftrightarrow$ 9, '(' $\leftrightarrow$ 11, ')' $\leftrightarrow$ 13.

More complicated formulas are mapped to more complicated numbers. The basic signs appear in this list sheathed in single quotation marks. They are being mentioned, and not used. Those sheaths serve as their name, and their naming takes place from beyond $F_{PA}$. It is an outside baptism.[16] Gödel numbering is ingenious because of the way it reverses the natural line of sight. The director's eye normally goes from signs to numbers. Gödel numbering encourages a reversal, one that goes from numbers to signs. The coordination achieved suggests a Cartesian coordinate system, but something stranger, too, and more difficult to grasp.

The prolegomenon to Gödel's proof consists of forty-six definitions. Each depends on the one that has come before, and each embodies a primitive recursive function.[17] These functions admit of a peculiar kind of definition. Both Richard Dedekind and Peano made use of the technique. The addition of any two natural numbers is resolved into two clauses. The first establishes that adding $x$ to zero goes nowhere beyond $x$: $0 + x = x$. The second defines addition in terms of succession and downward descent: $x + S(y) = S(x + y)$. Three plus four is the successor to three plus three. Downward descent follows: three plus three is the successor to three plus two. Descent continues until it achieves its appointed apotheosis in 0. Three plus four is the sevenfold successor of 0. Multiplication? The same. "As one can easily convince oneself," Gödel writes, "the functions $x + y$, $x \cdot y$, $x^y$ and furthermore the relations $x < y$ and $x = y$ are primitive recursive."[18]

Division does not figure in the Peano axioms.[19] Gödel's first definition brings it to life:

$$x/y =_{df} (\exists z)(z \leq x \ \& \ x = y. z).$$

There are no surprises. Whatever the number $x$, it is divisible by $y$ just in case there is a number $z$, less than or equal to $x$, such that $x$ is the product of $y$ and $z$. This is, after all, what division means. The sentence '$x/y =_{df} (\exists z)(z \leq x \ \& \ x = y. z)$'—the whole thing—is no part of $F_{PA}$. Still, it is possible to think of '$x/y$' as a derived sign, with '$(\exists z)(z \leq x \ \& \ x = y. z)$' indicating a path back to the basic signs of the system.

"We will now define a sequence of functions (relations)," Gödel goes on to write, "each of which is defined from the preceding ones."[20] The definitions accumulate, one after another, each following from the one that has gone before, and each admitting the discipline of definition by primitive recursion. The expanding sequence of definitions very quickly encompasses concepts such as *formula*, *axiom*, and *immediate consequence*. These are not obviously arithmetical. They are not arithmetical at all. Gödel numbering shows this earthy view to be primitive. Relationships between signs may be mirrored by relationships between numbers.

Forty-five definitions having been given; the forty-sixth, and last, defines the predicate '**Bew** $(x)$':

$$\mathbf{Bew}\ (x) =_{df} (\exists y)(y\ \mathbf{Bw}\ x).$$

The sign '**Bew** $(x)$,' this definition affirms, may be replaced by '$(\exists y)$ $(y\ \mathbf{Bw}\ x)$.' The sign '$y\ \mathbf{Bw}\ x$' is, in turn, defined by the forty-fifth definition, and so back to the first. The sign '**Bw**' is by birth arithmetical. In the strictest of strict senses, it stands for nothing beyond itself. But by virtue of its birth, it is intended to designate a relationship between numbers. At the same time, '**Bw**' enjoys a second interpretation by virtue of its place in logical society. It stands for the German *Beweis*, or proof. Under this interpretation, '$y\ \mathbf{Bw}\ x$' coordinates two numbers $y$ and $x$, where $y$ is the Gödel number of a proof of a formula whose Gödel number is $x$.

These definitions fix in amber or aspic the arithmetic predicate '**Bew** $(x)$' *and* the logical predicate '**Bew** $(x)$.' Peering into his own film, the director is now able to see himself peering into his own film.

A chain of definitions takes '**Bew**' to the heart of $F_{PA}$ and its basic signs. The definitions that precede the last are all primitive recursive. They depend on the definitions that have gone before until they empty themselves out at 0. The forty-sixth remains apart. It stakes a claim. It says that some $x$ is **Bew**. Its existential quantifier is not bounded.[21] It is the difference between a closed- and an open-ended search.

So far as primitive recursion goes, it is all the difference in the world.

## A Step Before the Last

There is a tight connection between Gödel's definitions and the formulas within $F_{PA}$. This is the subject of Gödel's fifth theorem. It is a theorem in two parts. Suppose that $R(n_1, \ldots, n_n)$ is a primitive recursive function or predicate controlling the numbers $n_1, \ldots, n_n$. The numbers, note—the real things. It follows that $R(n_1, \ldots, n_n)$ may be *represented* within $F_{PA}$ by a formula in which signs stand where vernacular variables stood: $\underline{R}(\underline{n_1}, \ldots, \underline{n_n})$. Underlining serves to underscore the distinction between signs *within* the formal system and symbols *about* the formal system or the natural numbers. If $R(n_1, \ldots, n_n)$ is a primitive recursive function, predicate, or relation, there is always a *translate* $\underline{R}(\underline{n_1}, \ldots, \underline{n_n})$ within $F_{PA}$ such that

1. if $R(n_1, \ldots, n_n)$ is true in N, then $\underline{R}(\underline{n_1}, \ldots, \underline{n_n})$ is *provable* within $F_{PA}$.
2. If $R(x_1, \ldots, x_n)$ is false—then not.

As logicians say,

$$\mathbb{N} \vDash R(n_1, \ldots, n_k) \Rightarrow F_{PA} \vdash \underline{R}(n_1, \ldots, n_k)$$
$$\mathbb{N} \vDash \sim R(n_1, \ldots, n_k) \Rightarrow F_{PA} \vdash \sim \underline{R}(n_1, \ldots, n_k).$$

This claim is compelling. If "2 + 2 = 4" is true in N, then its translate is provable within $F_{PA}$. If not, of what use $F_{PA}$? The theorem does not, by itself, identify the translate.[22] What can be said in its favor is that it exists. It is hard to imagine a tighter connection, but if the connection is tight, it is tight only for the primitive recursive functions. Proof is otherwise. It is not primitive recursive, satisfying 1. but not 2. The first forty-five of Gödel's predicates and relations are strongly represented within $F_{PA}$. The forty-sixth, and last, weakly represented.

And this, too, is something logicians say.

## Incompleteness

If Peano arithmetic is consistent, there is a formula within $F_{PA}$ such that neither it nor its negation is provable in $F_{PA}$.[23] The formula is undecidable. Peano arithmetic is incomplete. Gödel's proof of his great theorem is rebarbative—there is no other word. But in the opening paragraphs of his treatise, Gödel, writing with his own matchless concision and elegance, offers an accessible, if informal, account of his argument.

Consistency gives way in favor of the assumption that every provable formula is true. Suppose all the formulas of $F_{PA}$ with one free variable $x$ so arranged that the $n$th formula is designated $R(n)$. A variable $x$ is free if it does not lie within the scope of any quantifier, so that $F(x)$ says neither that everything is F nor that something is F. The symbol $[\alpha; n]$ designates from beyond $F_{PA}$ a formula deep in the bowels of $F_{PA}$. And more: it provides a recipe for its construction. It is the formula that results when $x$ is replaced by $\underline{n}$ in $\alpha$, the variable giving way to the numeral.[24]

With this said, consider the class $K$ of natural numbers such that

$$n \in K \equiv \sim\mathbf{Bew}[R(n); n].$$

The definition of $K$ makes use only of concepts that have already been defined. It is of a piece with the forty-six definitions on record. It follows that there is a formula $\mathbf{G}(n)$ in $F_{PA}$ that, if only it could talk, would say that $n \in K$. "There is not the slightest difficulty," Gödel observes, "in writing out the formula $\mathbf{G}$."[25]

The formula designated by $\mathbf{G}$ is *within* $F_{PA}$. Thus

$$\mathbf{G} = R(q)$$

for some number q, since the formulas of $F_{PA}$ have been gathered into a list in which everything is included and nothing is left out.

The conclusion of the incompleteness theorem is now budding on the bunched fingertips of this argument. $[R(q); q]$ is an instance of $[\alpha; n]$. The recipe enjoined by $[\alpha; n]$ is in force. $[R(q); q]$ is the formula within

$F_{PA}$ that results when $q$ replaces $x$ in $R(q)$. The Gödel number of $R(q)$ is $q$. Whence by a chain of definitions

$$G(q) \equiv [R(q); q] \equiv q \in K \equiv \sim\mathbf{Bew}[R(q); q].$$

$[R(q); q]$ is a mute collocation of signs. By itself, it says nothing, but like an ancient rune, when read rightly, it says that $R(q)$ cannot be demonstrated.[26] But $\mathbf{G} = R(q)$. $\mathbf{G}$ says as much. It says it of *itself*. And it *must* exist.

The argument proceeds by contradiction. If true, then by **I**, it must be provable.[27] But since $q \in K$ it must be *un*provable as well.

It cannot be both.

The assumption that $\mathbf{G}$ is unprovable in $F_{PA}$ leads again to contradiction. The argument, but not the proof, is over.

## Inexpressible Truth

At much the same time as Gödel published his incompleteness theorems, Alfred Tarski published his treatise about the concept of truth in formalized languages.[28] Gödel and Tarski seem to have anticipated one another, circumstances that themselves convey an air of paradox. Formal arithmetic, having served as a system's skeleton, must now acquire the musculature of real life—an interpretation in the natural numbers. Signs become symbols. An interpreted version of $F_{PA}$ has among its symbols any number of individual variables, the usual sentential connectives and quantifiers, and, at most, denumerably many predicate variables of various finite ranks. Some home assembly is yet required. By a model, logicians mean an ordered pair $\mathbf{M} = < D, \varphi >$, where $D$ is the domain of $\mathbf{M}$, and $\varphi$ a function assigning to the predicate variables of $F_{PA}$ relations of corresponding rank on $D$. An assignment $\alpha$ is a function that maps the individual variables of $F_{PA}$ onto individuals in $D$. Whence

$$\alpha \text{ satisfies } S(x, \dots) \text{ in } \mathbf{M},$$

defined by recursion on the length of $S$. A sentence is a formula with no free individual variables, and, as such, is either satisfied by every assignment or by none. It is True in $\mathbf{M}$, or False in $\mathbf{M}$. There is no wishy-washiness about it.

Tarski's argument is scorpion shaped. Its sting is in its tail. The first few steps are obvious. A property or relation is definable within **M** if and only if there is a formula in $F_{PA}$ defining it.[29] The number $x$ is even if it is the sum of two identical numbers. The requisite formula is '$\exists y(x = y + y)$.' A relation *in* **M** has been defined by a formula *of* $F_{PA}$. This is mildly marvelous, but no more than that.

Let Tr be the set of Gödel numbers of the true sentences of $F_{PA}$. Is Tr definable in **M**? If so, there must be a formula **Tr**$(x)$ in $F_{PA}$ such that **Tr**$(x)$ is satisfiable in **M** only for the interpretation of **Tr** as Tr. Suppose **Tr** defined in $F_{PA}$ so that for *every* sentence, there is a proof that **Tr**($\underline{S}$) $\leftrightarrow$ S, where $\underline{S}$ is the name of S. Whereupon there is ~**Tr**($\underline{S}$) $\leftrightarrow$ S, where $\underline{S}$ encodes the Gödel number of '~**Tr**(S).'[30]

There is no formula **Tr**. That is the sting.

## The Mind as a Machine

When Gödel's treatise first appeared in English, John Lucas concluded that men were not machines:[31]

> Gödel's theorem must apply to cybernetical machines, because it is of the essence of being a machine that it should be a concrete instantiation of a formal system. It follows that given any machine which is consistent and capable of doing simple arithmetic, there is a formula which it is incapable of producing as being true—i.e. the formula is unprovable-in-the-system—but which *we can see* [emphasis added] to be true. It follows that no machine can be a complete or adequate model of the mind, that minds are essentially different from machines.[32]

Roger Penrose revived the claim and endowed it with the luster of his great reputation. "Human understanding and insight," he argued, "cannot be reduced to any set of computational rules." This conclusion, Penrose insisted, followed from Gödel's theorem. "There must be more to human thinking," he added plaintively, "than can ever be achieved by a computer."[33] Because this thesis expressed a common human hope, or,

as much, a common human fear, it was rejected by a number of philosophers and logicians.

Very early on, Hilary Putnam argued that the Lucas-Penrose argument was flawed. What can legitimately be established by the incompleteness theorem is only the conclusion that

1.   $F_{PA}$ is consistent if and only if **G** is undecidable.

What is more, 1. may itself be demonstrated within $F_{PA}$. Absent a
proof of the consistency of $F_{PA}$, 1. does nothing to show anything.[34] And
by Gödel's second incompleteness theorem, there can be no proof of the
consistency of $F_{PA}$ within $F_{PA}$.[35]

No one, least of all Lucas or Penrose, considering them as two heads
on one body, should have argued the contrary. It is easy to see that **G**
must be true; difficult to see that $F_{PA}$ is consistent; and in neither case is
a proof forthcoming in $F_{PA}$. What can be demonstrated is 1.; what can
be seen is **G**. These are two different claims. They belong to two different
orders of thought.[36]

*On ne va tout de même pas pinailler pour si peu.*

Nor does 1. demonstrate anything more than an incidental kinship
between consistency and undecidability. They are provably equivalent
within $F_{PA}$. They are not generally the same. If they were, then any
consistent system would, by definition, be incomplete. Presburger arithmetic stands as a case to the contrary. Consistency and incompleteness
coincide for systems rich enough to generate whole number arithmetic.

Lumbering after Putnam, Panu Raatikainen has acquired his
shadow. The Lucas–Penrose argument is invalid, Raatikainen argues,
because "in general, we have no idea whether the Gödel sentence of an
arbitrary system is true. What we can know is only that the Gödel sentence of a system is true if and only if the system is consistent, and this
much is provable in the system itself."[37]

The first of these claims is false, the second, incoherent. No idea?
Gödel's proof contains a strong, sustained, and powerful argument that
**G** is not provable in $F_{PA}$. This is what **G** says. That should give anyone

some idea whether **G** is true. Were **G** false in N its negation would be true in N, and, in that case, demonstrable in F$_{PA}$. This is so plainly an unwholesome conclusion that no one is disposed to endorse it, least of all Gödel: "From the remark that [R(q); q] says about itself that it is not provable, it follows at once that [R(q); q] is true, for [R(q); q] is indeed unprovable (being undecidable)."[38]

There remains Raatikainen's claim that "the Gödel sentence of a system is true if and only if the system is consistent." So far as it goes, no one might scruple. That "this much is provable in the system itself" is otherwise. It provokes scruples all around. And for every good reason. There is no way to express the truth of **G** within F$_{PA}$.[39]

What has been neglected in this discussion is the reason that it began: the wish, or the need, to show that men are not machines. This conclusion remains as far from the premise of any argument as it ever was. We may allow Peano to become one with F$_{PA}$, and allow Lucas to see what Peano could not see and never saw. It hardly follows that Lucas may not be identified with a formal system, too, both men, like eighteenth-century courtiers, immured in formality. There is not the slightest reason, Alonzo Church once remarked to me, that the system used to describe F$_{PA}$ should not itself be formalized, resulting in some still more elaborate system, F$_{FPA}$, the superintendent of systems as much identified with F$_{FPA}$ as Peano is with F$_{PA}$. In his course in mathematical logic at Princeton University, Church endeavored to do just that. It was a tedious exercise, Church, glacial and self-contained, in the end lecturing to one or two students.

The phenomenon of incompleteness led Gödel to a characteristically subtle affirmation: "Either mathematics is incompletable in this sense, that its evident axioms can never be comprised in a finite rule, that is to say, the human mind (even within the realm of pure mathematics) infinitely surpasses the powers of any finite machine, or else there exist absolutely unsolvable Diophantine problems of the type specified."[40]

If this affirmation is subtle, it is also mysterious. A mathematician, in contemplatively considering F$_{PA}$, is in a position to see something that cannot be expressed within $F_{PA}$. Just how does this entail the conclusion that the human mind "infinitely surpasses the power of any finite machine?" The incompleteness theorems carry something like a poisoned seed. If $FF_{PA}$ is a formal system, it, too, is open to incompleteness, the truth of its Gödel sentence luridly shimmering from the perspective of still some further system. Far from indicating that the human mind is not machinelike, Gödel's argument shows only that nothing in the incompleteness theorems rules out the possibility that it is machines all the way up.

Gödel, who anticipated so much of the discussion, anticipated this as well: "Such a state of affairs would show that there is something nonmechanical in the sense that the overall plan for the historical development of machines is not mechanical. If the general plan is mechanical, then the whole [human] race can be summarized in one machine."[41]

Of this, all one can say is that either the general plan is mechanical or it is not, and no one knows which it is. The thesis that the human mind is a machine is returned to the same dark defile that has already entombed the study of consciousness. The defile is dark because there is nothing to be said, and it is a defile because everyone wishes to say it.

## The Director's Cut

Peano saw what he could see. We see in the light of the incompleteness theorems, and they have changed how things are seen. The truth is neither completely provable nor is it completely definable. These conclusions have retained their power to shock after ninety years. They were established at great expense. To establish these results, Gödel and Tarski needed to collapse multiple perspectives onto a single canvas, the mathematician looking at the natural numbers, the logician looking at the mathematician. But as in all great art, the incompleteness theorems demand, and as often receive, a still further perspective, the viewer beyond the canvas, the reader behind the proof, *his* vision, *his* point of view. The incompleteness theorems could not otherwise live in the minds of future generations. The process of withdrawal and encompassing reflection, if indefinite, is also endless. The incompleteness theorems provoke an intellectual experience only against the ever-receding background of a common language, a way of life, a culture of contemplation. No matter the degree to which things have been formalized, the formal objects that result are always described, and so discerned, amidst the inconvenience and distraction of the world in which we find ourselves explaining things to one another, specifying inferences, drawing conclusions, signing contracts, writing love letters and, in general, getting on with life in ways that are inseparably a part of life.

No formal system explains itself. It cannot say anything and we cannot say everything.[42]

If this is not a fact of logic, it is a fact of life, and so a feature of life. Every description of voluntary action is as incomplete as Peano arithmetic. No matter the network of causal influences acting on a human being, its description inevitably comes to seem incomplete, the strings in plain sight, the puppet master hidden.

There can be no proof of this, of course, but it is something we sense, and it is true. I do not know that we can expect anything more from life, or from logic.

There it is: $<P_A, \mathbb{N}>$, the director's cut. So it is. So we are. Even so.

# 22. Comments on Stuart Pivar's *Lifecode*

*Lifecode: The Theory of Biological Self-Organization*
By Stuart Pivar
Ryland Press, 164 pp.

*L*IFECODE IS AN AMBITIOUS AND PROVOCATIVE ATTEMPT TO DEAL with the problem of form in biology. It is ambitious because the nature and succession of biological forms, or *morphogenesis*, has been a subject of absorbing interest throughout the nineteenth and twentieth centuries, and it is provocative because it proposes a global solution to the problem of morphogenesis. The strength of this book lies in its detailed and remarkably beautiful illustrations. Its weakness is a matter of its disturbing lack of mathematical sophistication.

Pivar's chief thesis is that all biological forms are in essence variations of the torus. More specifically, Pivar argues, as much by implication as affirmation, that given a reticulated torus $T$ and its associated continuous deformations $D = d_1, d_2, \ldots, d_k$; and given a variety of biological structures $B = b_1, \ldots, b_n$, there is a mapping $G: D \to B$, such that $G$ is *form-preserving*.[1]

If successive deformations of the torus are form-preserving, Pivar argues, this suffices to explain morphogenetic change. Thus in a paper dealing with chicken embryology, Pivar argues that a "rational account for the forms of the familiar stages of chick embryogenesis" may be obtained "by demonstrating that the stages of the inversion of a reticulated toroidal surface can produce the same series of configurations."

Biologists will at once respond that if a distinguished series of deformations of a torus explains the successive changes in the form of a chick embryo, then so does a series of still photographs. Why not, in fact, claim that the sequence of morphological changes evident in chicken embryology explains *itself*, since plainly there is one mapping—the identity mapping—that satisfies Pivar's explanatory conditions? To duplicate biological structures at a distance by means of pictorial analogies is not yet to explain them. Pivar, it is true, very often writes as if the inversion of a reticulated torus represents some fundamental process in nature, but in *Lifecode* he neither expresses this hypothesis as a theorem nor demonstrates it as a fact.

In writing about the same issues in *Structural Stability and Morphogenesis*, René Thom observed that biologists find it difficult to believe "that we can construct an abstract, purely geometrical theory of morphogenesis, *independent of the substrate of forms, and the nature of the forces that create them.*"[2] Thom was quite correct. Biologists have found it difficult to believe this, and for good reason. It is not clear at all that forms in nature have a common origin and thus anything like a common explanation. A grapefruit and a baseball, after all, share the same form, but for entirely different reasons. An indifference to "the nature of the forces that create" morphological forms is not obviously an attitude calculated to improve our understanding.

Thom countered this criticism by arguing that under quite specific conditions, certain morphological forms appear necessarily in all structures falling under the same general description—very roughly, the space of all smooth maps between $n$-dimensional manifolds. He was then able to prove a strong classification theorem for the singularities of these maps. There is nothing comparable in Pivar's work. Biologists rejected Thom's analysis very largely because it was qualitative and not quantitative. Pivar's analysis is neither. It is instead *illustrative*, and while this may stimulate the imagination, it cannot yet be considered a scientific theory.

There is, in addition, some reason to suspect that taken simply as an illustrative account, Pivar's treatment raises as many questions as it might answer.

Pivar bases his analysis on the claim that all biological forms may be obtained by means of successive inversions of the reticulated torus. In Euclidean 3-space, the torus may be described parametrically by the equations $x = (c + a \cos v) \cos u$; $y = (c + a \cos v) \sin u$; and $y = (c + a \cos v)$ sin $u$. Inversions of a ring torus—Pivar's crucial model—yields either a ring cyclide or a parabolic ring cyclide.

Are these forms truly of crucial importance in theoretical biology even as illustrations? If so, why? To be sure, the flows of a Hamiltonian system may be embedded on the surface of an $n$-dimensional torus. But this interesting idea is not one that Pivar explores.

The role assigned to the torus in Pivar's text is for a different reason unpersuasive. A rectangle may be continuously deformed into a torus. If the torus in turn may be continuously deformed into something resembling the reticulated surface of a chick embryo, are we thus to say that the true fundamental form in nature is the rectangle? If not the rectangle, why the torus?

But by the same token, the torus cannot be continuously deformed into a sphere; and yet the sphere, appearing in such forms as the fertilized egg or the human skull, seems far more a fundamental shape in biology than the torus.

These questions reflect a deeper conceptual issue. The crucial notion of *form* itself is given no appropriate definition within Pivar's text. We thus do not know when two forms are the same and when they are different. The set $f: X \to Y$ of continuous mappings between topological spaces $X$ and $Y$ is in this regard too general to provide anything of interest to biology. But by the same token, the set $g: X \to Y$ of affine mappings between two geometrical spaces $X$ and $Y$ is too narrow to provide anything of interest to biology. It hardly helps in this regard that Pivar embraces both topological and geometrical models without seeming to realize that they have quite different properties and that great care and precision must be employed when topology and geometry are used in a single model.

Thus biologists can (and will) argue that a torus and rectangle do not have the same form, even though one may be deformed into the other, while topologists will (and can) argue that from a topological point of view, they do.

Who is to prevail?

A much greater difficulty in Pivar's theory arises when the all-important issue of the succession of forms is considered. Any account of either embryology or evolution must deal with the fact that morphogenesis involves stable discontinuities. Cell division in embryology is most obviously a process in which initially one cell becomes two cells. Going from one cell to two separate cells cannot be expressed as the result of a continuous transformation. To say that this process has been explained by observing that parent and daughter cells share a common, and perhaps even a generic form, is unhelpful inasmuch as it is so severely phenomenological as to provide nothing of interest beyond the obvious.

But Pivar's formal analysis, which is never stated explicitly but hinted at in the text, has no mathematical machinery by which discontinuous processes may effectively be analyzed. An ordinary rectangle may, for example, be deformed in various ways: it may be puckered, dimpled, crumpled, folded, or formed into a cusp. Of these deformations, only the fold and the cusp are structurally stable in the sense that small perturbations of their shape yield shapes close to the original. This is not true of the dimple.

To accommodate structurally stable deformations, a far richer and more complex analysis is required, one that involves the set of smooth maps between $n$-dimensional manifolds and the study of their singularities.

This is a subject studied by a number of great mathematicians: Marston Morse, Hasler Whitney, Stephen Smale, René Thom. The absence of their names from Pivar's bibliography is evidence that while

Pivar has very successfully appreciated the problem of morphogenesis, he has not—yet—been able to command the mathematical tools needed to analyze it.

# 23. CATASTROPHE THEORY AND ITS APPLICATIONS: A CRITICAL REVIEW

*Catastrophe Theory and Its Applications*
BY TIM POSTON AND IAN STEWART
PITMAN, 491 PP.

## Introduction

From a mathematical point of view, catastrophe theory is a contribution to the theory of the singularities of smooth maps; but it achieved prominence in virtue of its applications to biology and the social sciences. René Thom is the theory's creator, in large measure, and was its most persuasive advocate.

A Fields medalist, his own credentials as a modern mathematician were, of course, exquisite. But at some time in the early 1960s, he began to doubt his ability to maintain his own standards in pure mathematical research. He was a member of the Institut des Hautes Études Scientifiques at the moment that Alexander Grothendieck began his decade-long domination of algebraic geometry, and Grothendieck was also a member of the IHES. Thom was rueful but honest about his interactions with Grothendieck. He attended a few of his seminars, he told me, but was oppressed by Grothendieck's crushing technical superiority. The remark was characteristic of the man. I can think of no other great mathematician who would have said anything comparable.

Withdrawing from pure mathematics, Thom advanced into theoretical biology. He was able to keep the company of his new colleagues with all the confidence of a great mathematician, but none of the shrewdness of an adroit academic. "We have to let biologists busy themselves," he would later write, "with their very concrete—but almost meaningless—experiments; in developmental biology how could they hope to solve a problem they cannot even formulate?"[1] Skeptics recalled the dismal experiences of mathematicians in biology. Before the cracking of the genetic code by purely biochemical means, any number of capable mathematicians had glutted the journals with code-theoretic conjectures that in retrospect seem to revolve languidly around an axis signifying utter irrelevance.

In 1972, Thom cast his bread upon the waters with the publication of *Structural Stability and Morphogenesis*. Mathematicians reviewing the book agreed, almost to a man, that whatever its flaws, *Structural Stability and Morphogenesis* is a mature work of the imagination. If this was the book's first assessment, it will remain, I suspect, its last. With the exception of D'Arcy Thompson's *On Growth and Form*, there is no other book like it, and certainly no contemporary work in theoretical biology with the same mathematical elegance and sophistication.

Christopher Zeeman read Thom with great enthusiasm. At the Centre for Theoretical Physics in Trieste, where we were both lecturing, he remarked to me that he thought Thom's theories as important as the development of the calculus. In a series of widely read papers touching on biology, sociology, psychology, and political science, he championed the new theory. Zeeman was himself a distinguished and influential topologist. His voice carried conviction and it was heard. Zeeman's papers on catastrophe theory were issued under the title *Catastrophe Theory: Selected Papers: 1972–1977*.[2]

Virtually no one without a professional concern for the singularities of smooth maps actually read Thom's book. The material it presents is incomprehensible to the layman and inaccessible to the ordinary mathematician. The book's natural antagonists suspected that were they to avow roundly that *Structural Stability and Morphogenesis* contains nothing more than a strong string of solemn absurdities, the inevitable result would be exposure on grounds of technical incompetence.

It remained, then, for mathematicians to address catastrophe theory critically. Early reviewers of Thom's book (J. Guckenheimer, for example, writing in the *Bulletin of the American Mathematical Society*) had already observed, with just a touch of asperity, that what might loosely be called Thom's Program was by no means free of purely technical difficulties. H. J. Sussmann and R. Zahler, in their turn, addressed the applications of catastrophe theory to biology and the social sciences. Their conclusions were unusually harsh, at least by standards of polemics common in mathematics. "No CT (catastrophe theory) model that we have seen," they wrote, "is quantitatively correct, and the qualitative conclusions drawn are frequently wrong or vague or tautologous."[3] Sussmann and Zahler took every precaution not to attack Thom directly, at least not on purely mathematical grounds. Yet they managed to suggest that for all his undisputed greatness as a mathematician, Thom was confused, distinctly French in his combination of obscurity and prolixity.

Their criticism provoked a certain controversy within applied mathematics, a field not noted for the flamboyance of its exchanges. Zeeman's defenders were slow in coming to Thom's defense. Not being used to the high standards of indignation common in, say, literary criticism, they found themselves without an effective vocabulary of objurgation. The best they could muster was the claim that Sussmann and Zahler indulged in misrepresentation and misquotation. J. Guckenheimer went so far as to remark in a letter to *Nature* that Sussmann and Zahler were "snide."

## Differential Topology

Catastrophe theory sprawls like an irregular polyhedron over differential topology. Topology is a kind of loose-limbed geometry: continuous transformations predominate throughout. In differential topology, as in differential geometry, the methods of the calculus are brought into play.

The result, in Milnor's happy phrase, is "topology from a differential point of view."

Functions of several variables are treated, of course, and the spaces over which they range are more abstract than the plane. Much of the analysis is conceived in terms of manifolds. These are higher-dimensional analogues to an ordinary surface, but a differentiable curve is only a one-dimensional manifold when described by a mathematician with a taste for fancy generalities. Manifolds are, in any case, structures whose strengths are strictly local. For any point on a manifold of dimension $n$, there exists a neighborhood of that point homeomorphic to the interior of an $n$-dimensional Euclidean sphere. Between two manifolds there are continuous mappings, and, if a particular mapping and its inverse are both differentiable, one has a diffeomorphism. Manifolds and diffeomorphisms are what one gets in differential topology.

All this is quite abstract, but differential topology makes contact with concepts and concerns already present in the theory of ordinary differential equations. In one sense the theory of ordinary differential equations is dominated by computations. Equations suggest an unknown: Given $dx/dt = F(x,t)$ what function, if any, answers to $x$? For the most part, the answers cannot be derived analytically. The mathematician who actually takes it into his head to solve a differential equation must repair to the computer. The high point of theory comes about with the classical demonstration that solutions to such differential equations exist and are unique.

Ordinary differential equations also are geometrical objects. This was the discovery of Poincaré and Lyapunov. By a state of a system of ordinary differential equations $dx_i/dt = F_i(t,x_i)$, $i = 1,2, ...,n$, we mean the smallest set of numbers needed at $t_0$ to predict uniquely the course of the system at points beyond. Plainly, this number is $n + 1$. $R^{n+1}$ thus constitutes the system's phase space, and when $f$, in turn, defines a vector field on $R^{n+1}$, the pair together make up a dynamical system. Classical physics looked to such systems with the hope of actually solving the equations, this as a function of variations in the initial conditions in the system's states. Poincaré, however, stressed the global view. The entire set of trajectories defined by the integration of the vector field became an object of study. This leads to a portrait of the system's flows. If the underlying phase space is suitably abstract, on the order of a differentiable manifold, say, then the study of flows and the study of diffeomorphisms come tolerably close to one another since flows turn out to be nothing more than diffeomorphisms of a manifold onto itself.

For all that, the interpenetration of differential topology and the qualitative theory of ordinary differential equations was not completely obvious until quite recently. Poincaré had conceived differential equations in terms of flows or trajectories. However, a strong geometric spirit suggests strong geometric questions: What, for example, does it mean to say that one dynamic system is qualitatively like another? Precisely what topological structure does a family of ordinary differential equations induce on its phase space? Neither Poincaré nor Birkhoff, whose work on dynamic systems was especially profound, could answer these questions with complete generality. The introduction of topological structure in the qualitative theory of ordinary differential equations was a collaborative and cumulative effort by mathematicians of brilliance: Morse, Whitney, Milnor, Thom himself, Smale, Peixoto, and, more recently, Kupka, Williams, Anosov, and Arnold.

## Singularities of Smooth Maps

Catastrophe theory, as a mathematical theory, is most closely bound up with the study of singularities of smooth maps. Catastrophe theory can also be conceived in a mushier sense as a set of mathematical concepts. In still a looser sense, catastrophe theory is a general position in the philosophy of science, a collection of quasi-formal models in theoretical biology, an attitude, a point of view. Taking catastrophe theory in its narrowest sense means accepting the mathematician's creed that theories are where the theorems are.

One would think that any mathematical theorem, compactly expressed in only a finite number of words, could be explained in only a few more. To a certain extent this is true. Thom's theorem provides an elegant classification of the singularities of certain smooth maps until that point when they can no longer be classified. That having been said, one need only add that the theorem is both subtle and deep, an accomplishment in analysis comparable at least to the Atiyah-Singer index theorem. While Thom's name is associated with the theorem honorifically, a detailed proof was first composed by John Mather on the basis of work undertaken by Bernard Malgrange.

The space of all smooth mappings between manifolds makes up the most general context afforded by the theory of smooth maps. A classification is called for, but here the porousness of the problem makes a solution impossible. Local pathologies discretely dapple a space of stunning size. Where there is no pathology, there is rebarbative complexity instead. As a general strategy, differential topologists aim to study only the generic functions—most but not all. The generic functions form a subset at once open and dense, their complement of measure 0.

Morse theory constitutes the simplest case. A singular point of a function $f: R^n \to R$ is said to be nondegenerate if the matrix of second partial derivatives (Hessian) of $f$ evaluated at the singular point is invertible. On compact manifolds Morse theory provides an analysis of functions whose singular points are nondegenerate. The analysis is local throughout. If the singular points of a given function $f$ are nondegenerate, then the singular points of any function $g$ near $f$ also will be nondegenerate. The singularity structure of such functions is in a sense completely determined by degeneracy conditions on any of them. The ensemble forms a set on compact manifolds that is open and dense, and hence generic. With the exhibition and analysis of just this generic set of functions, the classification problem for smooth maps on compact manifolds is complete.

The general case is considerably more complicated. The basic object of study is yet again the space of smooth maps $F_i: U \to R$, where $U$ and $R$ are arbitrary manifolds, the $C^\infty$ topology prevailing throughout. Mather

(1973) studied this situation with notable success.[4] Stability considerations enter the picture and the generic functions turn on a choice of what Mather calls the nice range of dimensions. There the generic functions are those that are $C^\infty$ stable. For $C^0$ stability, the stable maps are generic for arbitrary manifolds $U$ and $R$, where $U$ is compact. This recalls the results of Morse theory. In the $C^\infty$ case there is an interesting and important play between stability and conditions of nondegeneracy. As one would expect, the stable maps are such that their singular points are distinct and nondegenerate.

There is Morse theory, then, and Mather's theory too. Sussmann organized the relationship between these areas in this way. Singularities of smooth maps are determined by two abstract objects: a dynamic system $(N,V)$ where $N$ as the most general phase space is a smooth manifold and $V$ is a vector field, and a map $F: N \to P$ where $P$ is itself a manifold. This mix of mappings on manifolds and dynamic systems suggests again the roots of catastrophe theory both in differential topology, where mappings and manifolds hold sway, and the more abstract reaches of the theory of ordinary differential equations, which is given over to dynamic systems broadly conceived. Morse theory, as well as Mather's theory, comes about when $V = 0$. What remains are the manifolds $N$ and $P$ and the maps between them. On the other hand, $V$ may be fully fledged. Families of functions now become crucial and individual functions skip over singularities—this as the result of degenerate singularities.[5]

Degenerate singularities, in fact, are a clue to the presence of catastrophes. Consider again the case of smooth mapping $f$, where $f$ is a real valued function. A catastrophe in $f$ is, roughly speaking, a kind of glorified singularity, which occurs when $f$ is embedded in a family of nearby functions, $H$, say. Catastrophes come about when the number of minima in $H$ change abruptly as parameters defining $H$ are varied. Thus the number of minima in $f_a = x^3 - ax$ changes dramatically as $a$ crosses from 1 to $-1$. Morse functions on compact manifolds are such

that any function near a given Morse function $f$ will exhibit nondegenerate singular points if $f$ does. This means that any function near $f$ has just
as many minima as $f$ itself. Functions with nondegenerate singularities
do not undergo catastrophe.

For a given function $f$ with degenerate critical points, the immersion of $f$ in a family of functions $F_c$, defined by some parameter $c$, is all
important. It is this immersion which makes such unstable functions
mathematically accessible. It is this immersion, too, that makes for a
mathematical model with some physical structure since the family of
functions comes about as the result of the influence of a vector field on
a manifold. Vector fields signify, in the end, differential equations. Differential equations tie the mathematical structure to some or another
physically realizable phenomena. The embeddings that figure most con-
spicuously in catastrophe theory Thom calls unfoldings. Let us call all
functions $g$ that agree locally with a given $f$ near a point $x$ the germ of
$f$. Following Callahan,[6] we can say that catastrophes are fixed by a pair
consisting of (i) the germ of a smooth real valued function $f$ taken at a
degenerate singular point, and (ii) the unfolding of $f$.

Catastrophe theory proper, taken as a mathematical creation, coin-
cides with Thom's theorem, and Thom's theorem is concerned with the
classification of singularities on smooth maps. The theorem's great appearance of complexity arises because the singularities occur in families
of functions parameterized by exogenous variables.

Suppose that $F$ denotes the entire space of $C^\infty$ functions on $R^{n+r}$, the
Whitney $C^\infty$ topology given again. We consider a smooth $r$-parameter
function of $n$ variables $f y_i, \dots, y_r (x_1, \dots, x_n)$, This way of writing things divides the parameters that figure in the function from the variables themselves. It would be just as easy to depict $f$ as a function in normal form:
$f: R^n \times R^r \to R$. Let $M_f$ be a manifold in $R^{n+r}$ defined by the conditions that
grad $f_x$ is zero, and consider strictly the catastrophe map $\chi_f : M_f^r \to R^r$
induced by the projection of $R^{n+r}$ downward onto $R^r$.

It is the catastrophe map that Thom's theorem treats. What it establishes is that if the codimension of $R^r$ is less than or equal to five, there

exists a generic $F^* \subseteq F$ such that for $F \in F^*$, where $M_f$ is a differentiable $r$-manifold embedded in $R^{n+r}$, the singularities in $\chi_f$ are all equivalent to one of a finite number of elementary catastrophes where the number of elementary catastrophes is an especially simple and intuitively attractive function of $r$. Up to $r = 5$ the singularities are finite. At points beyond, they are infinite. $\chi_f$ is itself stable in $M_f$.

Two unfamiliar technical terms figure in the statement of the theorem beyond the usual notions of differential topology. An equivalence is taken to be an equivalence under diffeomorphism. Singularities are topologically equivalent when the definition holds locally.

By a stable map $f:M \rightarrow N$ Thom means a map such that every map $g$ close to $f$ is topologically equivalent to $f$ in the space of all smooth maps between $M$ and $N$.

Generic is a fancy way of saying most but not all. This I have already mentioned, but in the present context genericity is ancillary to the concept of transversality or general position. Transversality I discuss below but for the moment it suffices to remark that generic functions predominate because the catastrophe map is itself transverse to the natural stratification of $F$.

The theorem shines through best in lower dimensions. Consider a smooth function $f$ on a one-dimensional space $\chi$ and imagine $f$ parameterized by just two parameters $a$ and $b$ drawn from a two-dimensional space $C$. Let $M$ be a surface fixed by the equilibrium values of $f$ determined when $\partial f / \partial f = 0$, $x$ a coordinate for $X$. It follows that $M$ is a smooth surface in $C \times X$. The only singularities in the projection of $M$ onto $C$ are fold curves and cusps. Actually, this conclusion holds for most infinitely differentiable functions $f$. The genericity of Thom's theorem carries downward.

Zeeman calls $C$ the control space and $X$ the behavioral space. Other writers prefer to talk of internal and external variables. In any event, when this specialization of Thom's theorem is introduced, the inevitable illustration features a view of $R^3$ as a pigeon might see it: the control space squared off below and the manifold itself draped over the middle ground which it divides into layers by means of a billowy pleat.

## Catastrophe Theory and Its Applications

As the title suggests, *Catastrophe Theory and Its Applications*, by Tim Poston and Ian Stewart (1978), is consecrated both to the exposition and application of Thom's work. This double purpose divides the book. Half the volume discusses the mathematical facts. The other half discusses the applications in physics, biology, and the social sciences.

Its high purposes notwithstanding, this is not a book for the mathematicians specializing, say, in the theory of semi-groups and afflicted with an unassuageable hunger for learning in differential topology. For them *Differentiable Germs and Catastrophes*[7] will be a more professional and hence a more sensible choice. Zeeman has recently compiled his own writings on catastrophe theory and mathematicians who need only the details to grasp the theory may find them set out in Zeeman's long but accurate proof of Thom's classification theorem. The first seven chapters of this book are not a good bet for the social scientist whose last encounter with abstract mathematics was a course on statistics.

Poston and Stewart make every effort to take the reader through the standard topics gently. Proofs are suppressed and definitions are made informal. However, progress from the early chapters, which set out the basics of linear and multilinear algebra and multidimensional geometry, to Chapter 7, where the concepts of an unfolding and determinacy make their much delayed appearance, is anything but smooth. The reader who manages to stay the course probably could have picked up the story with less effort.

The true audience for the book probably will be physicists and engineers, mathematically sophisticated social scientists, and perhaps philosophers. The mathematical development they will find in the first seven chapters of the book is correct and in many places pedagogically inspired. Poston and Stewart present their material in a grandfatherly tolerant fashion. The authors see catastrophe theory as an accretion on the calculus. The expansion of a function by means of a Taylor series is basic to their view. They use such expansions deftly both to explain the issues involved in the unfolding of a function and to provide a sense of continuity between classical and advanced points of view. Indeed, this stress on the historical development of catastrophe theory as a recognizable mathematical idiom is an especially attractive feature of the book's early chapters. There are many pictures throughout. Examples abound and the authors have taken pains to provide a running commentary on the mathematical analysis.

Still, much of the real mathematics is missing from this text. Elementary theorems are demonstrated, but the deep, pure, and powerful propositions upon which catastrophe theory itself rests, with the exception of the splitting lemma, are cited but not shown. Inevitably this makes for misunderstanding.

The mathematical account is not especially broad. Poston and Stewart see catastrophe theory in the context of the calculus, and this is all to the good. However, catastrophe theory is also bound up with general developments in differential topology, algebraic topology, and the qualitative theory of ordinary differential equations. There is no word in this volume of Smale or Peixoto, Mather's research is not mentioned, and no connection is drawn between the general topic of dynamic systems theory and catastrophe theory proper. Arnold and members of his Moscow school have contributed to the theory of the singularities of smooth maps. Except for assurances that his system of classification is too deep to be explained, his research makes no appearance in these pages either. Time is short, of course, and space shorter, but the reader who consults only this volume will come away with the conviction that catastrophe theory does not exist in a recognizable cultural setting.

## Philosophy and Catastrophe Theory

Catastrophe theory, when seen from afar, is an object of uncertain identity, something like swamp gas. From one perspective, there is the pure theory of elementary catastrophes, and this, as I have stressed, is a body of mathematical work. The chief results are Malgrange's $C^\infty$ version of the Weierstrass preparation theorem and Thom's classification theorem itself.

From another perspective, there is applied catastrophe theory, a subject of lurid efflorescence in biology, physics, and the social sciences. It is applied catastrophe theory that is discussed at great length in the second half of Poston and Stewart's volume: the geometry of fluids, optics, and scattering theory; the theory of elastic structures, where contact is made with the important work of Thompson and Hunt; thermodynamics and phase transitions, in which catastrophe theory is discussed in the context of renormalization methods; laser physics, and much besides, ranging from bee economics to alcoholism.

Thom's book is also an exercise in natural philosophy, a treatise, in fact, on such topics as structural stability, morphology, form and symmetry, and the meaning of explanation in science. It is a side of catastrophe theory that Poston and Stewart miss altogether. This omission compromises to a certain extent their claim to treat catastrophe theory as a unified subject.

Thom's book begins with a quotation from D'Arcy Thompson. Despite obvious differences in intellectual temperament and ability, Thom and Thompson are figures drawn from the same curve. Thom began his work on catastrophe theory as a speculative biologist puzzled, as was Thompson, by the mysteries of form. To see the world as a modern person sees it is already to suffer a certain intellectual corruption. The powerful and abstract concepts of contemporary physics are so much a part of our conceptual apparatus that many tangible and real processes evident to our senses now require an act of creative insight before they are noticed. The modern biologist, for the most part, looks on the organism dissectively. His ambition is not generally to understand the creature as an object fixed in a region of space and moving through time, but to get at its basic and more fundamental molecular structure.

This is reductionism in full flower and Thom properly describes it as a metaphysical position. However, biological systems are not only biochemistry made palpable. Organisms exist as geometric figures. Their lives can be traced as the sums of their cross-sections in time. Such arcs exhibit sharp discontinuities, of course, most notably at birth and death, but even these abrupt catastrophes are themselves part of a larger geometrical schema, an extended pattern of movement in space, something like a majestic periodic wave in any number of dimensions in which the perishable individual appears merely as a fragment in a figure.

Thom's objective is a theory of morphology or form. The generation and change of biological structures appears in such a theory as an aspect of geometry. The notion of form is connected to the concept of stability. That which has form is of necessity stable, Thom argues, and instabilities occur when symmetries are broken. Instabilities occur stably, too, so the breakdown of symmetrical structures reappears in nature, as changes in morphology. The forms themselves are dominated by their singularities since their singularity structure shapes their behavior. This suggests the influence of Riemann, who argued that what is important in a function are its singularities.

Physicists classify as phenomenological those theories that stay at the surface of things. Thom endorses the phenomenological point of view as a principle of philosophy. "That we can construct an abstract, purely geometrical theory of morphogenesis, *independent of the substrate of forms and the nature of the forces that created them,* might seem difficult to believe,"[8] Thom writes with a rare sense of his reader's potential skepticism. Just such a theory is what one gets in the pages that follow as Thom carries on a line of thought that he traces to Heraclitus.

A basketball and a grapefruit describe similar shapes in space. They are both spheres. On noticing that a grapefruit and a basketball are both round and have pebbly skins, the mathematician interested strictly in geometric structure might conclude that they are both nourishing and good to eat. To Sussmann this suggests what is wrong with phenomenological classifications, for he writes me that "science begins when superficial resemblance is dismissed as the main criterion for the classification of phenomena, and the analysis of structure and mechanism take over." This is the orthodox view, to be sure, and one with some plausibility. Yet what determines whether a particular resemblance is superficial? Even if grapefruits and basketballs have utterly different structures, why do they describe similar shapes in space?

These are not easy questions. Thom discusses them in difficult, incomplete chapters. There are many, many interesting remarks: on the primacy of geometric thinking; on reductionism; on the futility of global modeling; on the movement from local to global analysis, which Thom sees as a surrogate for the distinction between phenomenological and structural descriptions; on the concept of an object; and more.

When these chapters are integrated, the result is what Thom calls a general theory of models. Their qualitative behavior as abstract objects is meant to mimic, when boundary conditions are set, some of the local morphological properties of biological objects. The models that Thom presents are made up from chunks drawn from topology and the theory of ordinary differential equations. Thom assumes that the observable world is so arranged that to each point in time and space a state corresponds. The global evolution of a system of such states is determined by a vector field, which in effect fixes their flows. The set of states themselves come tolerably close to what physicists call a field. This makes for two fields, all in all, and some confusion. Vector fields are strictly mathematical objects. When the flow of a given system pushes at a representative point, objects situated in observable space may undergo discontinuous changes. Such are the catastrophes, and the global result is what Thom calls morphogenesis.

Thom's ultimate aim is to construct a theory that will synthesize purely local information into a global model that provides a description of the full space of observable objects, the vector field that runs it, and the set of catastrophes. The hypothesis of structural stability gives some specific content to his models, for Thom requires that the topology of the catastrophe set be stable with respect to perturbations of the vector field and the phase space itself. If this is a methodological principle, it is also something like a transcendental deduction.

## Structural Stability

Structural stability arises most naturally in the theory of differential equations. Andronov and Pontriagin discussed rough equations on the disk $D^2$. What they had in mind was the behavior of vector fields that stayed somehow the same when perturbed slightly. By perturbations they meant $C^1$ perturbations, and by "somehow the same" the existence of a homeomorphism $h$ that took trajectories of one vector field onto trajectories of another. Lefschetz purged the concept of its name and introduced in its place the modern sense of a structurally stable system. A vector field, or dynamic system $(X,V)$, is structurally stable just in case $(X,V+\partial V)$ is topologically close to $(X,V)$.

Closeness suggests a metric, so the concept of structural stability involves defining a suitable distance between vector fields or differential equations. In 1959 Peixoto considered the work initiated by Andonov and Pontriagin and showed that the structurally stable systems defined in the disk made up an open and dense set. Such is *genericity*, a concept that counts for much in the theory of ordinary differential equations, as well as differential topology, and a notion that, together with structural stability and transversality, forms an important part of the philosophic concepts Thom invokes. The fact that the structurally stable systems on $D^2$ are generic means, in effect, that most but not all such systems exhibit simple topological behavior. Even more to the point, genericity guarantees that the pathologies are all accessible. An unstable system may be approximated with whatever degree of fineness by a stable system so long as the stable systems themselves are open and dense.

Poston and Stewart introduce structural stability in a slightly different manner.[9] Their own examples are function theoretic, but they repeat Thom's philosophic arguments with regard to the importance of the concept:

> Let us return to the fundamental idea, structural stability; as we have said, the Universe is not chaos, and we can observe the recurrence of typical forms to which we give names. We must,

however, take account of the conditions of scientific observation itself. The experimenter cannot observe all the universe at once but, for his experiments and observations, is compelled to isolate a subsystem $S$ which is relatively independent of the rest of the universe. In practice he isolates and observes $S$ in a box $B$ whose geometric characteristics and the nature of whose attached measuring devices he specifies. Then he sets $S$ in a certain state $a$ which is defined by the preparation procedure, that is, the mode in which the box $B$ is filled, described as precisely as possible. Having set up state $a$, the experimenter observes or tests the contents of box $B$ sometime after the preparation of $a$. Each experimenter hopes that, if another experimenter performs this experiment at another time and place with box $B'$ obtained from $B$ by the action of an element of the Galilean group $G$ and with the same process of preparation of $a$, he will observe, to within experimental error, the same phenomena; without this hope, all would be in vain. However, no matter what precautions are taken to isolate $S$, the experimenter cannot remove entirely the interaction between $S$ and the outside world, and the conditions of a preparation procedure cannot be described and realized with perfect accuracy—and these initial differences cannot but perturb the evolution of the system. Therefore approximately equal results, i.e., equivalent under $G$, can be expected only after implicitly assuming that the evolution of $S$ from state $a$ is qualitatively stable, at least in respect to perturbations of the initial state and interaction with the outside world. In this way the hypothesis of structural stability of isolated scientific processes is implicit in all scientific observation.[10]

Thom is surely right in suggesting some notion such as structural stability for the analysis of what it means to be a scientific object. Yet there are objections to Thom's argument and at least some of them are mathematical. On a compact differentiable manifold $M^n$ a vector field $X$ is structurally stable if, given $\epsilon > 0$, there is a $\delta > 0$ such that whenever the distance between $X$ and $Y$ is less than $\delta$, $X$ and $Y$ are related by a homeomorphism $h$ within $\in$ of identity.

Let us call $\Sigma$ the set of structurally stable systems. For $n = 2, r \geq \mathbf{1}$, Peixoto[11] showed that $X \in \Sigma$ if and only if (a) singularities and closed orbits are generic; (b) saddle points remain unconnected by any trajectory; and (c) the $a$- and $\Omega$-limit sets of any trajectory are either a singular point or a closed orbit. What is more, systems that satisfy these conditions in $\Sigma$ are open and dense, hence generic. These conclusions make for an elegant and satisfying classification of the structurally stable systems by means of the formulation of equivalence classes taken with respect to homeomorphisms that preserve trajectories.

Life would surely be easier if Peixoto's results traveled upward. They do not. Smale generalized Peixoto's work by considering the space of all vector fields on a compact differentiable manifold $M^n$ with the $C^r$ topology. The example of the disk might have suggested that structural stability was crucial, but Smale aimed simply at the identification of a generic subset simple enough to suggest a schema for classification. Smale and Williams have shown that the structurally stable systems are *not* dense in higher dimensions, and hence not generic either. This suggests a lack of resonance between structural stability as purely a mathematical concept and structural stability as a philosophic condition. Thom's argument would have had a much greater intuitive plausibility if the structurally stable systems were dense everywhere. Isolated unstable systems could be discounted, the power of their pathologies remaining negligible.

Structural stability is by no means the only concept that can stand in for the informal notion of qualitative equivalence. The structurally stable systems are all the same under a topological transformation, but the underlying definition of equivalence may be altered. There is Smale's $\Omega$ stability, for example. We return, for illustration, to compact manifolds. Two functions $f$ and $g$ belonging to the diffeomorphisms of a compact $M$ are $\Omega$-conjugate if there is a conjugating homeomorphism between their nonwandering sets. $\Omega$-conjugacy is a finer and, hence, more subtle notion than topological equivalence. It leads naturally to an associated definition of stability. A given $f \in \text{Diff}(M)$, where $M$ is a compact manifold, is $\Omega$-stable if and only if there is a neighborhood of $U$ of $f$ in $\text{Diff}(M)$ such that for every $g$ in $U$, $g$ is $\Omega$-conjugate to $f$.[12] For all I know, there may be a dozen concepts similar in spirit but not in detail to Thom's notion of structural stability. Thom's own argument in favor of structural stability is philosophical and quite general. The definition of structural stability is mathematical and quite precise. Philosophers would be inclined to say that the second serves to explicate the first, but then they would wish to know why this mathematical concept is needed and not, say, $\Omega$-stability or something like it.

Finally, Thom's arguments in favor of structural stability raise a number of questions that lie somewhere between philosophy and mathematics. Some of these I have discussed at length elsewhere.[13] Others I have learned from Sussman. Structural stability is typically invoked to cover the case in which arbitrary perturbations might qualitatively destroy the recognizable characteristics of an object. Generally no distinction is drawn between those perturbations that might occur and those that are likely to occur. An object may well be structurally unstable and yet curiously persistent if the perturbations to which it is liable are improbable. Sussmann suggests the ordinary notion of shape as an example. The surface of my desk describes a rectangle in space and rectangularity is a structurally unstable property. The desk, however, is inelastic and stubbornly resistant to deformation. What might occur, does not; and the desk endures as just this desk despite the instability of its shape.

In a general way, Sussmann sees structural stability lining up with genericity as concepts Thom uses in the explication of form. To Sussmann, form suggests that which is nonrandom. In the Kolmogorov-Chaitin theory, sequences are random when no description shorter than a sequence length list is available. Randomness indicates informational sparseness, nonrandomness, redundancy, as in a pattern, or a musical theme. But nonrandomness or form and genericity or thickness are contrary concepts. What is generic is generally not highly specified. Thus in Sussmann's view, Thom has somehow busied himself with an analysis of formlessness or mere amorphousness, and like a baffled Beagle misled by a switch in scents has come crashing out of the underbrush with the concepts of structural stability and genericity in his dripping jaws.

I do not know how much of this I endorse. Sussmann himself seems to have achieved the same confusion by conflating physical and mathematical properties in his arguments against structural stability. The desk that sits by my side as I write surely instantiates the mathematical property of rectangularity. Its physical shape, the one that I notice when I tilt my head, is structurally stable and solidly resistant to perturbation. My desk remains recognizably the same in shape even though I nick it along its leading edge or even succumb to a vagrant impulse to scratch my initials across its surface. Those perturbations that Sussmann argues do not occur, do occur. We dismiss them because ordinary objects in space are seen as the same under small changes. However, this is a rough and ready account of structural stability, so perhaps Thom is right after all.

What seems more likely is that since they are defending contrary positions both may be wrong.

## Applications

Poston and Stewart spend just over two hundred pages on the applications of catastrophe theory. Here the cusp comes into its own and the reader gets to see the claims that have made catastrophe theory such a celebrated subject. The reach of this section of the book is vast, going as it does from laser physics to alcoholism. The chapters that make it up show some evidence of haste in composition because the number of solecisms and sheer grammatical bungles grow alarmingly once one pushes on past page 200. Nonetheless, each chapter deserves and will undoubtedly receive careful criticism.

My own reaction to this half of the book is that the authors manage to speak copiously of the applications of catastrophe theory only by making use of a threshold of success of markedly suspicious lowness. The trouble is that Poston and Stewart count as an application any use of catastrophe theory as a language and a set of organizing concepts.

Indeed, they are none too scrupulous about sticking to concepts that are manifestly catastrophe theoretic. All too often their argument in favor of catastrophe theory involves nothing more than the claim that known results can be given a new and perhaps more general formulation in the context of differential topology. Chapter 14, for example, is devoted to thermodynamics and phase transitions, topics that suggest functions that exhibit sharp discontinuities. Poston and Stewart do show that many of the famous old equations define a surface rather like a cusp once a change of variables has been defined. This establishes only the most marginal connections between thermodynamics and catastrophe theory, but at least there are concepts such as the cusp in common.

In Section 7, Poston and Stewart discuss Landau's theory of phase transitions. Here the argument turns on transversality or the concept of general position. In an intuitive sense transversality stands opposed to symmetry. Objects are in general position when they exhibit their typical or normal configuration, so transversality is a concept close to genericity as well.

The precise definition is given in a variety of contexts: vector fields, affine spaces or manifolds. Thus let $M$ and $N$ be smooth manifolds where $f:M \rightarrow N$ is a smooth mapping. Let $S$ be a submanifold of $N$ and say that $x \in M$. $f$ is transversal to $S$ at $x$ if and only if either (1) $f(x) \notin S$ or (2) $f(x) \in S$ and $T_{f(x)}S + (df)_x(T_xM) = T_{f(x)}Y$, where $T_xM$ is the tangent space to $M$ at $x$, and $(df)_x$ is the Jacobian of $f$ at $x$. Alternatively, I might have defined transversality strictly in terms of submanifolds and then set out what it is for a mapping to be transverse to a given submanifold. In any event, Poston and Stewart use transversality as a means by which certain arguments that figure physically in Landau's theory may be justified abstractly. With this tactic I have no real quarrel, but what in any of this is distinctly catastrophe theoretic is not clear.

Their use of transversality as a general concept in physics is another matter. At the end of their long chapter on thermodynamics and phase transitions Sir Arthur Eddington's shade is ceremoniously recalled: "In two dimensions any two lines are almost bound to meet sooner or later;

but in three dimensions and still more in four dimensions, two lines can and usually do miss another altogether, and the observation that they do meet is a genuine addition to knowledge."

Perhaps, but then again, perhaps not. In two dimensions one has a plane. Poston and Stewart have it that by transversality arguments pairs of line segments taken arbitrarily and extended indefinitely are likely to meet: "two manifolds taken at random," they write, "are infinitely unlikely to intersect nontransversely (in the same way that a real number chosen at random is infinitely unlikely to be exactly equal to $\pi$."[14] On the other hand, there are exactly as many pairs of parallel lines in the plane as pairs that intersect.

There is the appearance of confusion here, but no real paradox. There is still some way to go from the definition of transversality to its clear and successful employment as a principle in physics.

Catastrophe theory has had its successes, of course. Poston and Stewart present their most successful defense in Chapters 12 and 13, which are devoted first to optics and scattering theory, and then to the theory of elastic stability. Berry's important work is discussed and so is Thompson and Hunt's. By now this material has entered the scientific literature with intellectual assurance and it is perfectly reasonable to argue, as Berry does in a letter to *Nature*, that given these successes general arguments for the uselessness of catastrophe theory could not be correct. Readers of *Behavioral Science*, however, will probably be more closely interested in Chapter 17 of Poston and Stewart's book, which attacks what the authors call the problem of social modeling.

I had hoped to see Sussmann and Zahler come under strong countercriticism here, or failing that, in Poston and Stewart's short chapter on biology. Curiously enough, Sussmann and Zahler make only the briefest of appearances, and this in the bibliography. Poston has taken pains to rebut the critics of catastrophe theory, under the title "On Deducing the Presence of Catastrophes."[15]

Social life has long seemed especially resistant to mathematical analysis, and for obvious reasons. The plain evidence of our senses suggests processes that are strongly nonlinear, complex, discontinuous, and often chaotic. The conceptual elegance of Walrasian economics and the deep insights of game theory imply that this might be too superficial a view. Catastrophe theory has capitalized on the prevailing notion and achieved some tentative distinction as a mathematical discipline suited to modeling social and biological life. Thus many of the early applications of catastrophe theory were designed really to manage discontinuities, and for such purposes the Cusp catastrophe usually sufficed. Diagrams of the Cusp are in any case brilliantly informative, and even the sociologist who views mathematical symbolism as an affliction can tell with ease that a point traveling smoothly across a Cusp-like manifold is bound to jump at the edge of the pleat.

Thom read an even deeper significance into the Cusp. His basic line was quite plainly that the presence of a form such as the Cusp, or some similar catastrophe manifold in higher dimensions, *explained* the existence of discontinuities wherever they cropped up. This is phenomenalism taken twice over. What is important, in this view, is first the classification of phenomena on the basis of their surface similarities, and second the explanation of such similarities in terms of formally fixed mathematical creations, such as the Cusp or its analogue in higher dimensions.

Classical mathematics has any number of techniques for the management of discontinuity. Thom never suggests the contrary, although Zeeman sometimes writes as if mathematicians before coming upon the Cusp looked on any discontinuous phenomenon with rapt befuddlement. What Thom does celebrate is the unifying power of the Cusp. The veil of ignorance having been brushed away, one sees the same gently undulating manifold governing the snarls of agitated animals, the crash of stock markets, and the movements of militaristic societies.

It is this fondness for the Cusp that Sussmann and Zahler attack as unnatural. The problem, as they see it, is that embedded in the surface of the Cusp is a continuous family of thresholds. In Zeeman's model of animal aggression, for example, the behavioral surface describes the probability of an animal attack, given various combinations of rage and fear. All paths across the surface obey the delay rule. They jump from sheet to sheet only when a continuous continuation is no longer available.[16] Since the pleat itself vanishes, some paths obviously never cross a threshold at all. Animal attacks, Sussmann and Zahler observe, are by their nature strongly discontinuous. This implies that only those paths that actually jump across the pleat truly represent attacks. Thus a mathematician who manages to enrage but not intimidate a bull mastiff can confidently expect to keep out of harm's way, no matter what provocation he offers, since the path of this animal's behavior on the Cusp trots just above the pleat. This plainly suggests an empirical deficiency and the beginnings of a destructive dilemma.

Sussmann and Zahler object to the Cusp on stronger grounds. Sociologists have long known of shapes such as the cusp. Sharp discontinuities in social life must have suggested to them surfaces over which behavior jumps. What kept them from exhibiting such discoveries was a sense that the invocation of such a special shape requires something substantial by way of justification.

Zeeman seemed to provide what was missing, for in many of his papers, the particular choice of just this surface appeared as a consequence of Thom's classification theorem, which in Zeeman's view served to describe the shape of all possible equilibrium surfaces. The sociologist plotting bimodal behavior thus faced the Cusp as an inexpungable necessity, once he accepted the hypotheses of the theorem itself.

Sussmann and Zahler quite properly object to arguments along these lines. Even an informal statement of Thom's theorem, such as the one I outlined earlier, makes it plain that indefinitely many different surfaces are compatible with its conclusions. What the theorem does fix is their local singularity structure. However, the shape of the surface as a whole cannot be derived from the theorem itself. Of course, there are many plausible reasons why a scholar might favor a Cusp, a fondness for curves, say, or a predilection for pleats. Having made his choice, he can appeal to Thom's theorem for something sensible to say about its singularities. Unhappily, this suggests the activities of a researcher who simply invents the Cusp as a behavioral surface. The introduction of Thom's theorem was intended for purposes more dramatic than simply listing a set of singularities that in two dimensions are apparent on inspection.

On these issues, Poston and Stewart are silent.

## Conclusion

*Catastrophe Theory and Its Applications* is an important book. Many scholars will wish to own or read a copy. Curiously enough, despite its size and earnestness, it strikes me as remarkably less interesting than Thom's book.

The great trouble with Zeeman's models, for example, is simply that they are trivial. It hardly matters whether they are trivially true or trivially false, but *Structural Stability and Morphogenesis* suggests something quite different: the conceptual richness of much of modern mathematics. Philosophy stands in need of such concepts as structural stability, transversality, and genericity and so, indirectly, do the social sciences. Even if Thom is maddeningly imprecise, he is nonetheless interesting, and in the end that counts for almost everything.

# 24. Mathematics and Its Applications

## §0  Introduction

Group theory is about groups; the theory of rings, about rings. This suggests a generalization. Mathematics is about mathematics. Mathematics is about mathematics in the sense that entomology is about bugs. Who would deny it? If mathematics is about mathematics, it is about many other things as well. No one remarks on the unreasonable effectiveness of entomology. In counting two fingers and two fingers and reaching four fingers, numbers are being applied to fingers. In what does the application consist? The idea that there is a mapping between a subset of the natural numbers and the fingers of the human hand has all of the disadvantages of an arranged marriage. It appears reasonable only to those least involved in the proceedings. To the extent that the mapping is mathematical, it cannot have fingers in its range; and to the extent that it is not, it cannot have numbers in its domain.

Children nonetheless count their fingers with ease, and so do mathematicians.

How is it done?

In finger counting, we count fingers. This might suggest that numbers are one thing, fingers, another. But the creation of numbers, as Thierry of Chartres observed, is the creation of things. One finger is one finger necessarily. It could not be two fingers. If in counting fingers, we are mapping the number one to oneness, what remains of an *application* of numbers to things? One applies to oneness in just the sense in which one applies to itself. There is, after all, only one *one*. The statement that two fingers plus two fingers make four fingers is, when divided through by fingers, simply the statement that two plus two equals four. But one cannot divide through anything by fingers; and to leave the fingers out is only to return to the observation that two plus two equals four. This is nothing to sneeze at, of course, but it is nothing about which one might wish to write home.

What is unreasonably effective in mathematics is mathematics. What is unreasonably effective beyond mathematics is, as Eugene Wigner observed, a miracle.

## §1

Whatever its connection to other disciplines, mathematics frequently appears to be about itself, its concepts *self*-applied. Take groups. A set **G** closed under an associative binary operation **G X G → G** is a group if **G** includes an identity and an inverse. An identity **I** returns every element to itself: $a$ o **I** = $a$. An inverse returns any element to the identity: $a^{-1}$ o $a$ = **I**.

Groups are stable objects, important in mathematics as well as in physics. They play a role in topology, a subject devoted in large measure to the analysis of continuity. The most familiar topological space is the real line; its topology is defined by sets of open intervals. Now mathematicians, as well as philosophers, depend on the familiar upward movement of conceptual ascent, if only to rise above the smog and get a better view of things. A topological space is a particular item; this is the view from the ground. But the collection **Top** of topological spaces is a *category*; and this is the view that ascent reveals.

Categories are not simply sets. A set directs the mathematician's eye toward its elements $A$ and $B$; a category, toward the morphisms **Mor**($A$, $B$) between them. Morphisms may themselves be composed

$$\mathbf{Mor}(B, C) \; X \; \mathbf{Mor}(A, B) \to \mathbf{Mor}(A, C),$$

subject only to the triplet of conditions that:

**1.1**) **Mor**(*A*, *B*) and **Mor**(*A'*, *B'*) are either equal (*A* = *A'* and *B* = *B'*) or disjoint;

**1.2**) There is an identity morphism $id_A \in$ **Mor**(*A*, *A*) for every *A* in **Top**; and

**1.3**) Morphism composition is associative:

$$(h \circ g) \circ f = h \circ (g \circ f),$$

whenever $f \in$ **Mor**(*A*, *B*), $g \in$ **Mor**(*B*, *C*), and $h \in$ **Mor**(*C*, *D*)—this again for every *A*, *B*, *C*, and *D* in **Top**.

Wherein do groups figure? They figure in algebraic topology, a subject in which algebraic objects are assigned to topological structures in such a way that topological questions may be settled by algebraic methods.

Some definitions. A *path* or *arc* (or even a curve) in a topological space **X** is a continuous mapping *f* of a closed interval *I* = [*a*, *b*] into **X**. In what follows, *I* is always the closed interval [0, 1]. The images of *a* and *b* are the *endpoints* of the arc. A space is *arcwise connected* if any two points in **X** may be joined in an arc. Two paths in **X** are equivalent if one can be continuously transformed into the other.

Suppose that *f* and *g* are paths in **X** such that *g* starts where *f* ends. The product of *f* and *g* is:

$$(f \cdot g) = \begin{array}{ll} f(2t) & 0 \leq t \leq 1/2 \\ g(2t - 1) & 1/2 \leq t \leq 1 \end{array}.$$

The multiplication of paths is in general not associative; but associativity is recovered when paths are grouped into equivalence classes. It is easy now to define both an identity and an inverse. Assume this done.

A path is closed (or loops) if its initial and terminal points are the same. Let *x*, now, be any point of **X**. The set of all paths that loop from *x* to *x* satisfies group theoretic axioms; satisfying them, they form the *fundamental group* $\pi(\mathbf{X}, x)$ of **X** at the base point *x*.

A first connection between topology and algebra now emerges, like an image under darkroom developer:

**1.4)** If $X$ is arcwise connected, then $\pi(X, x)$ and $\pi(X, y)$ are isomorphic for any two points $x, y$ in $X$.

The theorem's contrapositive is somewhat more revealing: No matter the pair of points in $X$, if $\pi(X, x)$ and $\pi(X, y)$ are not isomorphic, $X$ is *not* arcwise connected. A topological condition has been determined by a group theoretic property.

Action at a distance.

As in all magic acts, one good trick suggests another. Let $S^1$, for example, be the unit circle in the real or complex plane. And let $f: I \to S^1$ be the closed path that goes around the circle just once:

$$f(t) = (\cos 2\pi t, \sin 2\pi t), 0 \leq t \leq 1.$$

$\xi(f)$ is the equivalence class of $f$. The obvious theorem follows:

**1.5)** The fundamental group $\pi(S^1, (1, 0))$ is an infinite cyclic group generated by $\xi(f)$.

The proof of **1.5**, like that of **1.4**, is a matter of applying diligently the definitions; but what follows is different, altogether dramatic. Let $E^n$ be the closed unit ball in Euclidean $n$-space; and let $f$ be a continuous map of $E^n$ into itself. Does $f$ have a point $x$ such that $f(x) = x$? It is a good question. The answer is provided by Brouwer's fixed-point theorem:

**1.6)** Any continuous map $f$ of $E^n$ into itself has at least one fixed point.

Unlike **1.5**, *this* theorem is an affirmation with a thousand faces, one of the protean declarations of mathematics.

The proof for $n \leq 2$ suggests the whole. First look to $n > 0$. A subset $A$ of a topological space $X$ is a *retract* of $X$ if there exists a continuous map $\gamma: X \to A$ such that $\gamma(a) = a$ for every $a$ in $A$. If $f: E^n \to E^n$ has no fixed points, then $S^{n-1}$ is a retract of $E^n$. Proceed by contraposition. $S^{n-1}$ is *not* a retract of $E^n$ when $n = 1$ because $E^1$ is connected, $S^0$, disconnected. Go, then, to $n = 2$. $\pi(S^1)$ is infinite cyclic; but $\pi(E^2)$ is a trivial group. It follows that $S^1$ is not a retract of $S^2$.

Categories were created with the aim of highlighting the mappings or morphisms between mathematical structures. The category **Top** of all topological spaces has already been fixed; ditto by definition the category **Grp** of all groups. A *functor* is a morphism between categories. If

*A* and *B* are categories their *covariant* functor *F*: *A* → *B* assigns to each object *a* in *A* an object *F*(*a*) in *B*; and assigns, moreover, to each morphism *f*: *A* → *A* a morphism *F*(*f*): *F*(*A*) → *F*(*A*).

The rules of the game are simple. For every *a* in *A*:

**1.7)** $F(id_A) = id(F_A)$,

where *id* is the identity morphism, and if *f*: *A* → *B* and *g*: *B* → *C* are two morphisms of *A*, then

**1.8)** $F(g \circ f) = F(g) \circ F(f)$.

*Contravariant* functors reverse the action of covariant functors; a pair, consisting of a covariant and contravariant functor, make up a *representation* functor.

Within algebraic topology, it is often useful (and sometimes necessary) to specify a topological space with respect to a distinguished point; such spaces constitute a category **Top\*** in their own right. A new category does nothing to change the definition of the fundamental group, of course, but it does make for a satisfying illustration of the way in which the fundamental group may acquire a secondary identity as a functor, one acting to map a category onto an algebraic object:

$$\pi(\mathbf{X}, x) : \mathbf{Top}^* \to \mathbf{Grp}.$$

This way of looking at things indicates that the fundamental group serves not simply to mirror the properties of a given topological space, but to mirror as well the continuous maps between spaces, a circumstance not all that easy to discern from the definition itself.

These considerations were prompted by a concern with mathematics self-applied. Herewith a provisional definition, one suggested by the functorial explication of the fundamental group. If *X* and *Y* are mathematical objects, then

**1.9)** **X** *applies* to **Y** if and only if there are categories *A* and *B*, such that **Y** belongs to *A* and **X** to *B*, and there exists a functor *F* on *A* such that *F*(**Y**) = **X**.

The language of categories and functors provides a subtle and elegant descriptive apparatus; still, categories and functors are mathematical objects and the applications so far scouted are internal to mathematics.

What of the Great Beyond? The scheme that I have been pursuing suggests that mathematics may be applied to mathematical objects; it makes no provisions for applications elsewhere.

## §2

A mathematical theory with empirical content, Charles Parsons has written, "takes the form of supposing that there is a system of *actual* objects and relations that is an instance of a structure that can be characterized mathematically." These are not lapidary words.[1] They raise the right question, but by means of the wrong distinction. They misleadingly suggest, those words, that mathematical objects *without* empirical content are somehow not actual. Not actual? But surely not potential either? And if neither actual nor potential, in what sense would mathematical theories without empirical content be about anything at all? The word that Parsons wants is *physical*; and the intended contrast is the familiar one, mathematical objects or structures on the one side, and physical objects or structures on the other.[2]

The question Parsons raises about mathematics, he does not answer explicitly, his discussion trailing off irresolutely. W. V. O. Quine is more forthright. "Take groups," he writes:

> In the redundant style of current model theory, a group would be said to be an ordered pair $(K, f)$ of a set $K$ and a binary function $f$ over $K$ fulfilling the familiar group axioms. More economically, let us identify the group simply with $f$, since $K$ is specifiable as the range of $f$. Each group, then, is a function $f$ fulfilling three axioms. Each group is a class $f$ of ordered triples, perhaps a set or perhaps an ultimate class.... Furthermore, $f$ need not be a pure class, for some of its ordered triples may contain individuals or impure sets. This happens when the group axioms are said to be *applied* somewhere in natural science. Such application consists in specifying some particular function $f$, in natural science, that fulfills the group axioms and consists of ordered triples of bodies or other objects of natural science.[3]

Whatever else they may affirm, these elegant remarks convey the impression that mathematical concepts (or predicates) are polyvalent in applying indifferently to mathematical *and* physical objects. "Group-hood," Quine writes (on the same page), "is a mathematical property of various mathematical and non-mathematical functions." This is rather like saying that *cowhood is a zoological property of various zoological and non-zoological herbivores*. If there are no non-zoological cows, why assume that there are some non-mathematical groups?

Skepticism about the application of mathematics arises as the result of the suspicion that nothing short of a mathematical object will ever satisfy a mathematical predicate. It is a suspicion that admits of elaboration in terms of an old-fashioned argument.[4] Let me bring the logician's formal language L into the discussion; ditto his set $K$ of mathematical structures. A structure's *theory* $T(K)$ is the set of sentences $\varphi$ of L such that $\varphi$ holds in every model $M \in K$. Let K constitute the finite groups and $T(K)$ the set of sentences true in each and every finite group. Sentences in $T(K)$, the logician might say, are *distributed* to the finite groups.

Nothing esoteric is at issue in this definitional circle. Distribution is the pattern when an ordinary predicate takes a divided reference. The logician's art is not required in order to discern that whatever is true of elephants in general is necessarily true of Bruno here. It is a pattern that fractures in obvious geometric cases. Thus consider *shape*, one of the crucial concepts by which we organize sensuous experience, and the subject, at least in part, of classical Euclidean geometry.[5] Is it possible to distribute the truths of Euclidean geometry to the shapes of ordinary life—desktops, basketballs, football fields, computer consoles, mirrors, and the like?

Not obviously.

In many cases, the predicates of Euclidean geometry just miss covering the objects, surfaces, and shapes that are familiar features of experience. Euclidean rectangles are, for example, bounded by the line segments joining their vertices. Rectangles in the real world may well be finite but *un*bounded, with no recognizable sides at all because beveled at their edges.[6] The chalk marks indicating the length and width of a football field have a determinate thickness and so contain multiple boundaries if they contain any boundary at all. Euclidean rectangles are structurally unstable: small deformations destroy their geometrical character. Not so physical rectangles. Such regions of space are robust. They remain rectangular and not some other shape despite an assortment of nicks, chips, and assorted scratches. The sum of the interior angles of a Euclidean rectangle is precisely 360 degrees; the interior angles on my desk sum to more or less 360 degrees.

*More or less.*

These particular cases may be enveloped by a general argument. It is a theorem that up to isomorphism there exist only two continuous metric geometries. The first is Euclidean, the second, hyperbolic. The categorical model for Euclidean geometry is the field of real numbers. It follows thus that if Euclidean geometry is distributed, physical space must locally be isomorphic to $\mathbb{R}^n$.[7] It is difficult to understand how the axioms of continuity could hold for physical points;[8] difficult again to imagine a one-to-one correspondence between physical points and the real numbers.[9] How would a correspondence between the real numbers and a variety of physical points be established?

*Experimentally?*

# §3

If distribution lapses in the case of crucial mathematical and physical shapes, it is often not by much, a circumstance that should provoke more by way of puzzlement than it does. The sum of the interior angles of a Euclidean triangle is precisely one hundred and eighty degrees—$\pi$ radians, to switch to the notation of the calculus, and then simply $\pi$, to keep the discussion focused on numbers and not units.[10] This is a theorem of Euclidean geometry, a fact revealed by pure thought. Yet mensuration in the real world reveals much the same thing among shapes vaguely triangular: the sum of their interior angles appears to follow a regular distribution around the number $\pi$. The better the measurement, the closer to π the result. This would seem to suggest a way forward in the project of making sense of a mathematical application. Letting M(Δ) and P(Δ) variably denote mathematical and physical triangles, and letting niceties of notation for the moment drift, **3.1** follows as a provisional definition, one that casts a mathematical application as the inverse of an approximation.

**3.1)** M(Δ) *applies* to P(Δ) if and only if P(Δ) is an approximation of M(Δ).

Now approximation is a large, a general concept, and one that appears throughout the sciences.[11] It is a concept that has a precise mathematical echo. Let E be a subset of the line. A point ξ is a *limit point* of E if every neighborhood of ξ contains a point $q \neq ξ$ such that $q \in E$. The definition immediately makes accessible a connection between the existence of a limit point and the presence of convergent sequences, thus tying together a number and a process:

**3.2)** If $E \subseteq \mathbb{R}$ then ξ is a limit point of E if and only if there exists a set of distinct points $S = \{x_n\}$ in E itself such that $\lim_{n \to \infty} \{x_n\} = ξ$.

Approximation as an activity suggests something getting close and then closer to something else, as when a police artist by a series of adroit adjustments refines a sketch of the perpetrator, each stroke bringing the finished picture closer to some remembered standard. **3.2** reproduces within point-set topology the connection between some fixed something and the numbers that are tending toward it, **S** and ξ acting as approximation and approximatee. The reproduction improves on the general concept inasmuch as convergence brings within the orbit of approximation oscillating processes—those governed by familiar functions such as $f(x) = x \sin 1/x$. So long as the discussion remains entirely within the charmed circle of purely and distinctively mathematical concepts, what results is both clear and useful. The Weierstrass approximation theorem serves as an illustration:

**3.3)** If f is a complex valued function, one continuous on [a, b], there exists a sequence of polynomials $P_n$ such that $\lim_{n \to \infty} P_n(x) = f(x)$ uniformly on [a, b].

The proof is easy enough, considering the power and weight of the theorem. It is surprising that *any* complex and continuous function may be approximated by a sequence of polynomial functions on a closed and bounded interval. More to the point, **3.3** gives to approximation a precise, independent, and intuitively satisfying interpretation.

Difficulties arise when this scheme is off-loaded from its purely mathematical setting. Mensuration, I have said, yields a set of numbers, but beyond any of the specific numbers, there is the larger space of possible points in which they are embedded—$\Omega$, say. Specific measurements comprise a set of points $S^*$ within $\Omega$. Relativized to the case at hand, the requisite relationship of approximation would seem now to follow:

**3.4)** $P(\Delta)$ is an approximation of $M(\Delta)$ if and only if $\pi$ is a limit point of $\Omega$.[12]

**3.4** is assumed even in the most elementary applications of the calculus. Talking of velocity, the authors of one text assert that "just as we approximate the slope of the tangent line by calculating the slope of the secant line, we can approximate the velocity [of a moving object] at $t = 1$ by calculating the average velocity over a small interval $[1, 1 + \Delta t]$." Approximate? Approximate how? By allowing average speeds to pass to the limit, of course, the answer of analytic mechanics since the seventeenth century.

But the usefulness of **3.4** is entirely cautionary. **3.5** follows from **3.4** and **3.2**:

**3.5)** $P(\Delta)$ is an approximation of $M(\Delta)$ if and only if there exists a set of distinct points $S$ in $\Omega$ itself such that $\lim\limits_{n \to \infty} \{x_n\} = \pi$.

And yet **3.5** is plainly gibberish. The real world makes available only finitely many measurements and these expressed as rational or computable real numbers. There exists no set of distinct points in $S$ converging to $\pi$ or to anything else. $\Omega$ is a subset of the rational numbers and the definitions of point-set topology are unavailing. Taken literally, **3.5** if true implies that $P(\Delta)$ is not—it is *never*—an approximation of $M(\Delta)$, however close points in $S^*$ may actually come to $\pi$. **3.5** must be taken loosely, then, but if **S** and **S**\* are distinct—and how else to construe the requisite looseness?—**3.1** lapses into irrelevance.

## §4

Symmetry is a property with many incarnations, and so a question arises as to the relationship between its mathematical and physical instances. It is group theory, Hermann Weyl affirms, that provides a language adequate to its definition.[13] Weyl's little book contains many examples and represents a significant attempt to demonstrate that certain algebraic objects have a direct, a natural, application to ordinary objects and their properties.

Let $\Gamma$ be the set of all points of some figure in the plane. A permutation on $\Gamma$ is a bijection $\Gamma \to \Gamma$; a given permutation $f$ is a symmetry of $\Gamma$ or an automorphism when $f$ acts to preserve distances. The set of all symmetries on $\Gamma$ under functional composition (one function mounting another) constitutes the group of symmetries $\mathbf{G}(\Gamma)$ of $\Gamma$.

Let $\Gamma$, for example, be the set of points on the perimeter of an equilateral triangle. Three sides make for three symmetries by counterclockwise rotation through 120, 240, and 360 degrees. These symmetries may be denoted as $R$, $R \circ R$, and $R \circ R \circ R$, which yields the identity and returns things to their starting position. There are, in addition, three symmetries $D_1, D_2,$ and $D_3$ that arise by reflecting altitudes through the three vertices of the triangle. The transformations

$$\Delta_3 = \{R, R_2, R_3, D_1, D_2, D_3\}$$

describe all possible permutations of the vertices of the given triangle. These being determined, so, too, are the relevant automorphisms.

So, too, the symmetric group $\Delta_3$.

A set **S** of symmetrically related objects is fashioned when a sequence of automorphisms is specified, as in **4.1**:

$$\begin{array}{ccc} A_1 & A_2 & A_k \end{array}$$

4.1)    $\Gamma \to \Gamma \to \Gamma ... \to ...\Gamma.$

The objects thus generated form a *symmetrical sequence* **S**. This suggests the obvious definition, the one in fact favored by Weyl:

**4.2)** $G(\Gamma)$ *applies* to **S** if and only if **S** is symmetrical on $\Gamma$.

So far, let me observe skeptically, there has been no escape from a circle of mathematical objects. Whatever the invigoration group theory affords, the satisfaction is entirely a matter of internal combustion. $G(\Gamma)$ is plainly a mathematical object; but in view of **4.1**, so, too, is **S**.

Nonetheless, an extended sense of application might be contrived—by an appeal to the transitivity of application, if nothing else—if sequences such as **S** themselves apply to sequences of real objects; applying directly to **S**, $\Gamma$ would then apply *in*directly to whatever **S** is itself applied. Thus

**4.3)** $G(\Gamma)$ applies to **S\*** if and only if **S** applies to **S\*** and $G(\Gamma)$ applies to **S**.

It is to **S\*** that one must look for physical applications.

And it is at this point that any definitional crispness that **4.3** affords begins to sag. The examples to which Weyl appeals are artistic rather than physical; but his case stands or falls on the plausibility of his analysis and owes little to the choice of examples. Symmetries occur in the graphic arts most obviously when a figure, or motif, occurs again and again, either in the plane or in a more complicated space. They exist, those figures or inscriptions or palmettes—the last, Weyl's example—in space, each separate from the other, each vibrant and alive, or not, depending on the artist's skill. But in looking at a symmetrical sequence of *this* sort, **4.1** gives entirely the wrong impression. The problem is again one of distribution and confinement. **4.1** represents a symmetrical sequence generated by $k$ operations on a *single* abstract object—$\Gamma$, as it happens. Those Persian bowmen or Greek palmettes or temple inscriptions are not generated by operations on a single figure. They are not generated at all. Each of $n$ items is created independently and each is distinct.[14] And none is quite identical to any other.

A far more natural representation of their relationship is afforded by mappings between spaces as in

**4.4)** $X \xrightarrow{f} Y \xrightarrow{g} Z \ldots \xrightarrow{h} \ldots W.$

If a connection to geometry is required, X, Y, Z, and W may be imagined as point sets, similar each to $\Gamma$: $f$, $g$, and $h$ take one space to the next. Functional composition extends the range of the mappings: $f \circ h = f\colon X \to W$. The sense in which **4.4** represents a symmetrical sequence may be expressed quite without group theory at all. Thus let the various functions be bijections; let them, too, preserve distances, so that if $x, y \in X$

$$D(x, y) = D(f(x), f(y)).$$

Each function then expresses an isomorphism. Congruence comes to be the commanding concept, one indicating that adjacent figures share a precisely defined similarity in structure.

But if **4.1** informs **4.4**, so that the sense of symmetry exhibited at **4.4** appears as group theoretical, it is necessary plainly that the following diagram must commute:

$$
\begin{array}{cccc}
A_1 & A_2 & A_k & \\
\Gamma \to \Gamma \to \Gamma & \dots \to & \dots \Gamma. \\
\downarrow & & & \downarrow \\
X \to Y \to Z & \dots \to & \dots W. \\
f & g & h &
\end{array}
$$

4.5)

when $\Gamma = X$.

And obvious, just as plainly, that it never does in virtue of the fact that $X \neq Y \neq Z \dots \neq \dots W$.

# §5

If groups do not quite capture a suitable sense of symmetry, it is possible that weaker mathematical structures might. A *semigroup* is a set of objects on which an associative binary operation has been defined. No inverse exists; no identity element is required. The semigroups have considerably less structure than the groups.[15] Functional composition is itself an associative operation. Say now that $S[X, Y, ..., Z]$ is any finite sequence of the point sets (or figures) X, Y, ... , Z, and let C be a collection of isomorphic mappings over **S**. **4.1**, **4.2**, and **4.3** have their ascending analogs in **5.1**, **5.2**, and **5.3**:

**5.1)** Isomorphisms over C form a *semigroup* **SG** under composition;

**5.2)** A sequence S[X, Y, ..., Z] is symmetrical in X, Y, ..., Z if X ≈ Y for every pair of elements X, Y in S[X, Y, ..., Z]; and

**5.3)** **SG** applies to **S** if and only if **S\*** is symmetrical in X, Y, ..., Z.

Transitivity of application is again invoked to fashion a notion of indirect application. The application of **SG** at **5.3** makes for a weak form of algebraic animation; but it does little to dilute the overall discomfort prompting this discussion. Let me reconvey my argument. **5.1** is an abstract entity, a sequence of point sets or spaces. The symmetries seized upon by the senses obtain between palpable and concrete physical objects. Symmetries thus discerned are approximate; the discerning eye does what it does within a certain margin of error. To the extent that a refined judgment of symmetry hinges on a definition of congruence, the distances invoked by the definition are preserved only to a greater or lesser extent. Thus if $x$ and $y$ are points in $X$, and $f: X \to Y$, then

$$D(x, y) = D(f(x), f(y)) \pm \delta.$$

At **5.1**, distances are preserved precisely.[16] If we had a convincing analysis of approximation, an analysis of applicability might well follow. One rather suspects that to pin one's hopes on approximation is in this case a maneuver destined simply to displace the fog backward.

Sections §3 - §5 were intended provisionally to answer the question whether a group takes physical instances. Asked provisionally, that question now gets a provisional answer:

No.

# §6

The arguments given suggest that nothing short of a mathematical object is apt to satisfy a mathematical predicate; it is hardly surprising, then, that within physics at least nothing short of a mathematical object *does* satisfy a mathematical predicate.

The systems theorist John Casti has argued that mathematical modeling is essentially a relationship between a natural system **N** and a formal system **F**.[17] Passage between the two is by means of an *encoding* map

$\zeta$: **N** $\rightarrow$ **F**, which serves to associate observables in **N** with items in **F**. The idea of an encoding map is not without its uses. The encoding map, if it exists at all, conveys a natural into a mathematical world: $\zeta$: **N** $\rightarrow$ **M**. For my purposes, the map's interest lies less with its ability to convey natural into mathematical objects, but in the reverse. If an encoding map exists, its *inverse* $\zeta *$: **M** $\rightarrow$ N should serve to demarcate at least one sense in which mathematical objects receive an application.

The argument now turns on choices for $\zeta *$. Within particle physics (but in other areas as well), **M** is taken as a group, **N** as its representative, and $\zeta *$ understood to be the action of a group homomorphism. Such is the broad outline of group representation theory. Does this scheme provide a satisfactory sense of application? Doubts arise for the simplest of reasons. If $\zeta *$ does represent the action of a group homomorphism, surely **N** is for that reason a mathematical object? If this is so, the scheme under consideration has in a certain sense overshot the mark, the application map, if it is given content, establishing that every target in its range is a mathematical object.

Consider a single particle—an electron, say—on a one-dimensional lattice; the lattice spacing is $b$.[18] The dynamics of this system are governed by the Hamiltonian

**6.1)** $H = p^2/2m + V(x)$,

where $m$ measures the mass of the electron and $p$ its momentum. $V$ is a potential function and satisfies the condition that:

**6.2)** $V(x + nb) = V(x)$

for every integer $n$. The system that results is *symmetrical* in the sense that translations $x \rightarrow x' = x + nb$ leave **H** unchanged. Insofar as they are governed by a Hamiltonian, any two systems thus related behave in the same way.

A few reminders. Within quantum theory, information is carried by *state vectors*. These are objects that provide an instantaneous perspective on a system, a kind of snapshot. Let Q be the set of such vectors, and let /y > and /y' > be state vectors related by a translation. The correspondence /|y> $\rightarrow$ /|y' > may itself be expressed by a linear operator T in Q:

**6.3)** $/y> \rightarrow /|y'> = T/y>$,

—this for every state vector $/y>$.

Not *any* linear operator suffices, of course; physical observables in quantum theory are expressed as scalar products $<F/y>$ of the various state vectors. It is here that old-fashioned numbers make an appearance and so preserve a connection between quantum theory and a world of fact and measurement. Unitary linear operators preserve scalar products; they preserve, as well, the length and angle between vectors. To the extent that **6.3** takes physically significant vectors to physically significant vectors, those operators must be unitary; so too the target of T.

Let us step backward for a moment. Here is the Big Picture. On the left there is a symmetrical something; on the right, another something. Symmetrical somethings of any sort suggest group theory, and, in fact, the set of symmetry operations on a lattice may be described by the discrete translation group $\mathbf{G}(T^D)$. Those other somethings comprise the set $\{T\}$ of unitary linear operators. $\{T\}$ constitutes a *representation* of the symmetry operations on **H**; it resembles an inner voice in harmony.

Now for a close-up. First, there is the induction of group theoretic structure on the alien territory of a set $L$ of linear operators in a vector space Q. $L$ becomes a group $\mathbf{G}(L)$ under the definition of the product of two operators $A$ and $B$ in $L$:

**6.4)** $Cx = A(B(x))$.

The identity is the unit operator. And every operator is presumed to have an inverse.

Next an arbitrary group **G** makes an appearance. The homomorphism

**6.5)** $h: \mathbf{G} \rightarrow \mathbf{G}(L)$

acts to establish a representation of **G** in $\mathbf{G}(L)$, with $\mathbf{G}(L)$ its representative. In general, group theory in physics proceeds by means of the group representation.[19] In the example given, $\mathbf{G}(TD)$ corresponds to **G**; $\{T\}$ to $L$; and given an $h$ such that

**6.6)** $h: \mathbf{G}(T^D) \rightarrow \mathbf{G}\{T\}$,

$\mathbf{G}\{T\}$ corresponds to $\mathbf{G}(L)$.

**6.5** has an ancillary purpose: it serves to specify the *applications* of group theory to physics in a large and general way. **6.6** makes the specification yet more specific. The results are philosophically discouraging (although not surprising). **G** and **G**(*L*) are mathematical objects; but then, so are **G**($T^D$) and **G**{T}. The real (or natural) world intrudes into this system only via the scalars. If there is any point at which mathematics is applied directly to anything at all, it is here. But these applications involve only counting and measurement. This is by no means trivial, but it does suggest that the encoding map carries information of a severely limited kind. An application of group theory to physics, on the evidence of **6.5** and **6.6**, is not yet an application of group theory to anything physical: so far as group representation goes, the target of every mathematical relation is itself mathematical; and as far as quantum theory goes, those objects marked as physical by the theory—the *range* of Casti's encoding map—do not appear as the targets of any sophisticated mathematical application.

This conclusion admits of ratification in the most familiar of physical applications. Consider the continuous rotation group **S0(2)**. The generator *J* of this group, it is easy to demonstrate, is $R(\phi) = e^{-i\phi J}$, where $R(\phi)$ is, of course, a continuous measure of rotation through a given angle. **S0(2)** is a Lie group; its structure is determined by group operations on *J* near the identity. So too its representations. Thus consider a representation of **S0(2)** defined in a finite dimensional vector space *V*. $R(\phi)$ and *J* both have associated operators $R(\phi)^*$ and $J^*$ in *V*. Under certain circumstances $J^*$ may be understood as an angular momentum operator in the sense of quantum mechanics. This lends to *J* a certain physical palpability. Nonetheless, $J^*$ is and remains an operator in *V*, purely a mathematical object in purely a mathematical space. The same point may be made about the actions of **SU(2)** and **SU(3)**, groups that play a crucial role in particle physics. **SU(2)**, for example, is represented in a two-dimensional abstract isospin space. The neutron and the proton are regarded as the isospin up and down components of a single nucleon. **SU(2)** defines the invariance of the strong interaction to rotations in this space. But **SU(2)** is a mathematical object; so, too, *its* representative. Wherever the escape from a circle of mathematical concepts is made, it is surely not here.

## §7

It is within mathematical physics that mathematics is most obviously applied and applied moreover with spectacular success, the very terms of description—*mathematical* physics—suggesting one discipline piggy-backed upon another. Still, to say that within quantum field theory or cosmology, mathematics has been a great success is hardly to pass beyond the obvious. A success in virtue of *what?* The temptation is strong to affirm that successes in mathematical physics arise owing to the application of mathematical to physical objects or structures, but plainly this is to begin an unattractive circle. It was this circumstance, no doubt, that prompted Eugene Wigner to remark that the successes of mathematical physics were an example of the "unreasonable effectiveness of mathematics."[20]

The canonical instruments of description within mathematical physics are ordinary or partial differential equations. In a well-known passage, David Hilbert and Richard Courant asked under what conditions "a mathematical problem... corresponds to physical reality." By a "problem" they meant an equation or system of equations. Their answer was that a system of differential equations corresponds to the physical world if unique solutions to the equations exist, and that, moreover, those solutions depend continuously on variations in their initial conditions.[21] Existence and uniqueness are self-evident requirements; the demand that solutions vary continuously with variations in their initial conditions is a concession to the vagaries of measurement:

> The third requirement... is necessary if the mathematical formulation is to describe observable natural phenomena. Data in nature cannot possibly be conceived as rigidly fixed; the mere process of measuring them involves small errors.... Therefore, a mathematical problem cannot be considered as realistically corresponding to a physical phenomenon unless a variation of the given data in a sufficiently small range leads to an arbitrary small change in the solution.[22]

Following Hadamard, Hilbert and Courant call a system of equations satisfying these three constraints *well posed*.

Well-posed problems in analysis answer to a precise set of mathematical conditions. Consider a system of ordinary first-order differential equations expressed in vector matrix form:

7.1) $dx/dt = f(x, t)$, $x(a) = b$.

Existence, uniqueness, and continuity depend on constraints imposed on $f(x, t)$. Assume thus that R is a region in $< x, t >$. $f(x, t)$ is *Lipschitz continuous* in R just in case there exists a constant $k > 0$ such that:

$$f(x_1, t_1) - f(x_2, t_2) \leq k \,|x_1 - x_2|.$$

Here $(x_1, t)$ and $(x_2, t)$ are points in R.

Assume, further, that $f$ is continuous *and* Lipschitz continuous in R; and let $\delta$ be a number such that $0 < \delta < 1/k$. And assume, finally, that $u$ and $v$ are solutions of **7.1**. Uniqueness now follows, but only for a sufficiently small interval (the interval, in fact, determined by $\delta$):

7.2) If $u$ and $v$ are defined on the interval $|t - t_0| \leq \delta$, and if $u(t_0) = v(t_0)$,

then $u = v$.

What of existence? Let $u_1, u_2, ..., u_n$ be successive approximations to **7.1** in the sense that

$$u_0(t) = x_0$$

$$u_{k+1}(t) = x_0 + \int_{t_0}^{t} f\left(x, u_k(t)\right) dt, \quad k = 0, 1, 2, ...$$

Suppose now $f$ is continuous in R: $|x - x_0| \leq a$, $|t - t_0| \leq b$, $(a, b) > 0$; and suppose, too, that $M$ is a constant such $f(t, x) < M$ for all $(t, x)$ in R. Let I be the interval $|x - x_0| \leq h$, where $h$ is the minimum of $\{a, b/M\}$. The Cauchy-Peano theorem affirms that:

7.3) The approximations $u_1, ..., u_n$ converge on I to a solution **u** of **7.1**.

7.3 is purely a *local* theorem: it says that solutions exist near a given point; it says nothing whatsoever about wherever else they may exist. The theorem is carefully hedged. And for good reason. There are simple differential equations for which an embarrassing number of solutions exist, or none at all. The equation $y^2 + x^2 y' = 0$ is an example. Infinitely many solutions satisfy the initial condition $y(0) = 0$. No solution satisfies the initial condition $y(0) = b$, if $b \neq 0$. At $< 0, 1 >$, this equation fails of continuity.

The Cauchy-Peano theorem does not apply.

7.3 may be supplanted by a global existence theorem, but only if $f$ is Lipschitz continuous for *every* $t$ in an interval I. It follows then that successive solutions are defined over I itself.

There remains the matter of continuity. Let $u$ be a solution of 7.1 passing through the point $(t_0, x_0)$; and let $u^*$ be a solution passing through $(t_0', x_0')$. Both $u$ and $u^*$ pass through those points in R. The requisite conclusion follows, preceded by a double condition:

7.4) If for every $\in > 0$, there is a $\delta > 0$ such that $u$ and $u^*$ exist on a common interval $a < t < b$; and if $a < t' < b$, $|t - t'| < \delta$, $|t_0 - t_0'| < \delta$, $|x_0 - x_0'| < \delta$, then $|u(t) - u(t)^*| < \in$.

To the extent that 7.1 satisfies the hypotheses of 7.2, 7.3, and 7.4, to *that* extent 7.1 is well posed.

The concept of a well-posed problem in analysis is interesting insofar as it specifies conditions that are necessary for applicability; but however necessary, they are, those conditions, hardly sufficient. How could they be? Like any other equation, a differential equation expresses an affirmation: some unknown function answers to certain conditions. The Cauchy-Peano theorem establishes that for a certain class $\Phi$ of differential equations, a suitable function exists. The elements of the theory $T(\Phi)$ satisfied in models of $T(\Phi)$ are true simply in virtue of being elements of $T(\Phi)$.

But true of *what*? Surely not physical objects? This would provide an access to the real world too easy to be of any use.

Like so many other mathematical objects, a differential equation is dedicated to the double life. If the first is a matter of the solutions specified by the equation, the second involves the induction of form over space. The simple differential equation:

7.5) $df/dt = Af(t)$

provides a familiar example. The association established between $t$ and $f(t)$ creates a coordinate system. The set of points evoked—$t$, on the one hand, $f(t)$, on the other—is the *phase* (or state) *space* of the equation. $Ax$ evidently plays $A$ against each of its phase points. This implies that $Ax$ represents the rate of change of $x$ at $t$. Rates of change evoke tangent lines and slopes. To each point in the $< t, f(t) >$ plane, a differential equation—*the* differential equation—assigns a short line segment of fixed slope. A phase space upon which such lines have been impressed is a *direction* or *lineal* field.

Imagine now that the plane has been filled with curves tangent at each point to the lines determining a lineal field. The set of such curves fills out the plane. And each curve defines a differential solution $u(t) = x$ to a differential equation, for plainly $du(t)/dt = Au(t)$ in virtue of the way in which those curves have been defined.

It is thus that a differential equation elegantly enters the geometrical scene. Nothing much changes in the interpretation of **7.1** itself. The lineal field passes to a vector field, but the induction of geometrical structure on an underlying space proceeds apace. The system of equations

7.6) $\quad dX/dt = y$
$\qquad dY/dt = -x$

assigns to every point in the $< X, Y >$ plane a vector $<y, -x>$. The trajectory or *flow* of a differential equation corresponds to the graph of those points in the plane whose velocity at $< x, y >$ is simply $< y, -x >$. Trajectories have a strange and dreamy mathematical life of their own. The Poincaré-Bendixson theorem establishes, for example, that a bounded semi-trajectory tends either to a singular point or spirals onto a simple closed curve.

An autonomous system of $n$ differential equations is one in which time has dwindled and disappeared. Points in space are $n$-dimensional, with $R^n$ itself the collection of such points. On $R^n$ things are seen everywhere as Euclid saw them: $R^n$ is an $n$-dimensional Euclidean vector space. On a differential manifold, things are seen locally as Euclid saw them; globally, functions and mappings may be overtaken by weirdness. Thus the modern definition of a dynamical system as a pair $< \mathbf{M}, \mathbf{V} >$

consisting of a differential manifold **M** and its associated vector field **V**. To every differential equation there corresponds a dynamical system.

On the other hand, suppose that one starts at a point $x$ of **M**. Let $g^t(x)$ denote the state of the system at $t$. For every $t$, $g$ defines a mapping $g^t: \mathbf{M} \rightarrow \mathbf{M}$ of **M** onto itself. Suppose that $g^{t+s} = g^t g^s$. Suppose, too, that $g^0$ is an identity element. That family of mappings emerges as a one parameter group of transformations. A set **M** together with an appropriate group of transformations now acquires a substantial identity as a flow or dynamical system. In the obvious case, **M** is a differential manifold, $g$ a differential mapping. Every dynamical system, so defined, corresponds to a differential equation. Since $g^t$ is differentiable, there is—there must be—a function $v$ such that $dg^t/dt = v(g^t)$. A differential equation may thus be understood as the action of a one parameter group of diffeomorphisms on a differential manifold.

The attempt to assess the applicability of differential equations by looking toward their geometric interpretation runs into a familiar difficulty. Neither direction fields nor manifolds are anything other than mathematical structures. And the association between groups and differential equations suggests that insofar as the applicability of differential equations must be defended in terms of the applicability of groups, the result is apt to be nugatory. There is yet no clear and compelling sense in which groups are applicable.

This may well suggest that the applicability of differential equations turns not on the equations, but on their solutions instead. An icy but invigorating jet now follows. The majority of differential equations cannot be solved by analytic methods. Nor by any other methods. This is a conclusion that physicists will find reassuring. Most differential equations cannot be solved; and they cannot solve most differential equations.

# §8

The exception to this depressing diagnosis arises in the case of *linear* differential equations; such systems, V. I. Arnold remarks, constitutes "the

only large class of differential equations for which there exists a defini-
tive theory."[23]

These are encouraging words.

Linear algebra is the province of linear mappings, relations, combinations, and spaces. Instead of numbers, linear algebra trades often in
vectors, curiously rootless objects, occurring in physics as arrows taking
off from the origins of a coordinate system, and in the calculus books as
directed line segments. In fact, vectors are polyvalent, described now in
one formulation, now in another, the various definitions all equivalent
in the end.

If $x_1, x_2, \dots, x_n \in R^n$, with $\{c_i\}$ a set of scalars, dying to be attached, the
vector $c_1 x_1 + \dots + c_k x_k$ is a *linear combination* of the vectors $x_1, x_2, \dots, x_k$—
linear because only the operations of scalar multiplication and vector ad-
dition are involved, a combination because something is being heaped
together, as in a Caesar salad. A set of vectors $\{x_1, x_2, \dots, x_k\}$ is independent if $c_1 x_1 + \dots + c_k x_k = 0$ implies that $c_1 = c_2 = \dots = c_k = 0$; otherwise,
dependent, the language itself suggesting that among the dependent vec-
tors, some are superfluous and may be expressed as a linear combination
of those that remain.

This makes for an important theorem, one serving to endow vec-
tors with an exoskeleton. Suppose that $S \subset R^n$ and let E be the set of all
linear combinations of vectors in S. Then S is said to be spanned by E.
This is a definition. It follows that E is the intersection of all subspaces
in $R^n$ containing S. Theorem and definition may, of course, be reversed.
A basis B of a vector space $X$ is an independent subset of $X$ spanning $X$.
This, too, is a definition.

Herewith that important theorem:

**8.1)** If B is a basis of $X$, then every **x** in $X$ may be uniquely expressed as
a linear combination of base vectors.

The representation, note, exists in virtue of the fact that B spans $X$.
It is unique in virtue of the fact that B is independent. Thus in $R^2$, every
vector $x = \langle x_1, x_2 \rangle$ exists first in its own right, and second as a linear

combination of basis vectors $x = x_1 e_1 + x_2 e_2$. The unit vectors $e_1 = <1, 0>$ and $e_2 = <0, 1>$ constitute the standard basis for $R^2$.

This is all to the good, but whether it is all for the best is another question.

The theory in which Arnold has expressed his confidence constitutes a meditation on two differential equations. The first is an inhomogeneous second-order linear differential equation:

**8.2)** $d^2 x/dx + a_1(t) \, dx/dt + a_2(t)x = b(t)$;

the second, a reduced version of the first:

**8.3)** $d^2 x/dx + a_1(t) \, dx/dt + a_2(t)x = 0$.

It is the play between these equations that induces order throughout their solution space. Let $L(x)$ denote $x'' + a_1(t)x' + a_2(t)x$ so that **8.2** is $L(x) = b(t)$. That order is in evidence in the following theorems; and these comprise the theory to which Arnold refers.

For any twice differential functions $u_k$ and constants $c_k$

**8.4)** $L(c_1 u_1(t) + c_2 u_2(t) + ... + c_m u_m(t)) = c_1 L(u_1(t) + c_2 L(u_2(t) + ...$
$+ c_m L(u_m(t)$.

**8.4** follows directly from the fact that **8.2** is linear.

**8.5)** If $u$ and $w$ are solutions to **8.2**, then $u - w$ is a solution to **8.3**.

$L(u(t) = b(t) = L(w(t))$. But then $L[u(t) - w(t)] = L(u(t)) - L(w(t))$
$= b(t) - b(t) = 0$.

**8.6)** If $u$ is a solution of **8.2**, and $w$ a solution of **8.3**, then $u + w$ is a
solution of **8.3**.

Say that $L(u(t)) = b(t)$; say too that $L(w(t)) = 0$. $L\{u(t) + w(t)\} = L(u(t))$
$+ L(w(t)) = b(t) + 0 = b(t)$.

Now let $u$ be *any* solution of **8.3**:

**8.7)** Every solution of **8.3** is of the form $u + w$, where $w$ is a solution of
**8.4**.

Let $v$ be a solution to **8.3**. By **8.5** it follows that $v - u = w$, where $w$ is a solution of **8.4**.

The *general* solution of **8.3** may thus be determined by first fixing a single solution $v$ to **8.3**, and then allowing $w$ to range over all solutions to **8.4**, an interesting example in mathematics of a tail wagging a dog.

Suppose now that $u$ is a solution of **8.4**:

**8.8)** If $u(t_0) = 0$ for some $t_0$ and $u'(t_0)$ as well, then $u(t) = 0$.

But $u(t) = 0$ is already a solution of **8.4**.

**8.9)** If $u_1, ..., u_m$, are solutions of **8.4**, then so is any linear combination of $u_1, ..., u_m$.

This follows at once from **8.7**.

**8.10)** If $u_1$ and $u_2$ are linearly independent solutions of **8.4**, then every solution is a linear combination of $u_1$ and $u_2$.

Note that **8.9** affirms only that linear combination of solutions to **8.4** are solutions to **8.4**; but **8.10**, that *all* solutions of **8.4** are linear combinations of two linearly independent solutions to **8.4**.

Let $u$ be any solution of **8.4**. The system **E** of simultaneous equations

$$u_1(0)x_1 + u_2(0)x_2 = u(0)$$
$$u_1'(0)x_1 + u_2'(0)x_2 = u'(0)$$

has a non-vanishing determinant. **E** thus has a unique solution $x_1 = c_1$ and $x_2 = c_2$. If $v(t)$ is $c_1u_1(t) + c_2u_2(t)$ it follows that $v(t)$ is a solution of **8.4** because it is a linear combination of solutions to **8. 4**. But $v(0) = u(0)$ and $v'(0) = u'(0)$; thus $v = u$ by the existence and uniqueness theorem for **8.2**.

It is **8.10** that provides the algebraic shell for the theory of linear differential equations, *all* solutions of **8.4** emerging as linear combinations of $u_1$ and $u_2$, and thus forming a *basis* for **S**, which is now revealed to have the structure of a finite dimensional vector space. This in turn implies the correlative conclusion that solutions of **8.3** are all of the form $w + c_1u_1(t) + c_2u_2(t)$, where $u_1$ and $u_2$ are linearly independent solutions of **8.4**.

A retrospective is now in order. The foregoing was prompted by the desire to see or sense the spot at which a differential equation or system of equations applies to anything beyond a mathematical structure. Well-posed differential equations are useful in the sense that if their solutions did not exist or were not unique, they would not be useful at all. Continuity is less obvious a condition, whatever the justification offered by Hilbert and Courant. Still, there is nothing in the idea of a well-posed problem in analysis, or a well-posed system of equations, that does more than indicate what physically relevant systems must have. What they do have, and how they have it, this remains unstated, unexamined, and unexplored. The qualitative theory of ordinary differential equations has the welcome effect of turning the mathematician's attention from their solutions taken one at a time to all of them at once. The imagination is thus enlarged by a new prospect, but the rich and intriguing geometrical structures so revealed does little, and, in fact, it does nothing, to explain the coordination between equations and the facts to which they are so often applied. So far as the linear differential equations go, V. I. Arnold is correct. There is a theory, and a theory, moreover, that has a stabilizing effect across the complete range of linear differential equations. This is no little thing. But while the theory draws a connection between linear equations and linear algebra, so far as their applications go, the connection is internal to mathematics, falling well within the categorical definitions of section §1.

The spot at which a differential equation or system of equations applies to anything beyond a mathematical structure?

I'm just asking.

# §9

Consider a physical system **P**. A continuously varying physical parameter $\xi$ is given, one subject intermittently to measurement, so that $g(t) = \xi$ is a record of how much, or how little, there is of $\xi$ at $t$. Or in the case of bacteria, how many of them there are. If the pair $<\mathbf{P}, \xi>$ makes for a physical system, there is by analogy the pair $<\mathbf{D}, f>$, where $\mathbf{D}$ is a differential equation, and $f$ its solution.

Let us suppose that for some finite spectrum of values, $f(t_k) \cong g(t_k)$.

The example that follows is the very stuff of textbooks. Having made an appearance at **7.5**, the equation

**9.1)** $df/dt = Af(t)$

is simple enough to suggest that its applications must be transparent if any applications are transparent. Solutions are exponential: $x = f(t) = Ke^{At}$. The pair $<\mathbf{D}, f>$, where $\mathbf{D}$ just is **9.1** and $f$ its solution, makes for a differential model.

Can we not say over an obvious range of cases, such as birth rates or the growth of compound interest, that there is a very accessible sense of applicability to be had in the play between differential equations and the physical processes to which they apply? Let me just resolve both $g(t) = \xi$ and $f(t) = Ke^{At}$ to $t = 0$ so that for the mathematician, **9.1** appears as an initial value problem, and for the biologist or the bacteria, as the beginning of the experiment. The voice of common sense now chips in to claim that for $t = 0$, and for some finite spectrum of values thereafter,

**9.2)** <D, f> *applies to* <P, ξ> if and only if $f(t_k) \cong g(t_k)$.

I do not see how **9.2** could be faulted if only because the relationship in which $< \mathbf{D}, f > $ *applies to* $< \mathbf{P}, \xi >$ is loose enough to encompass indefinitely many variants. One might as well say that the two structures are coordinated, connected, or otherwise companionably joined. If one might as well say any of that, one might as well say that differential equations are very often useful and leave matters without saying why.

But **9.2** raises the same reservations about the applicability of mathematics as considerably more complicated cases. It is a one-man multitude. If the inner structure of $< \mathbf{D}, f >$ and $< \mathbf{P}, \xi >$ were better aligned, one could replace an unclear sense of applicability by a mathematical mapping or morphism between them. Far from being well-aligned, these objects are not aligned at all. The function $g(t) = \xi$ is neither differentiable nor continuous; barely literate, in what respect does it have anything to do with $< \mathbf{D}, f >$? The function $f$, on the other hand, is differentiable and thus *continuous*. Continuous functions take intermediate values; in what sense does it have anything to do with $< \mathbf{P}, \xi >$? There is no warm point of connectivity between them. Differential and physical structures are radically unalike.

In that case, why should $f(t_k) \cong g(t_k)$?

We are by now traveling in all the old familiar circles.

## §10

"To specify a physical theory," Michael Atiyah writes in the course of a discussion of quantum field theory, "the usual procedure is to define a

Lagrangian or action $L$." A Lagrangian $L(\varphi)$ having been given, where $\varphi$ is a scalar field, the partition function $P$ of the theory is described by a Feynman functional integral. "These Feynman integrals," Atiyah writes with some understatement, "are not very well defined mathematically."[24]

The parts of the theory that *are* mathematically well defined are described by the axioms for topological quantum field theory. A topological **QFT** in dimension $d$ is identified with a functor $Z$ such that $Z$ assigns i) a finite dimensional complex vector space $Z(\Sigma)$ to every compact oriented smooth d-dimensional manifold $\Sigma$; and ii) a vector $Z(Y) \in Z(\Sigma)$ to each compact oriented $(d + 1)$ dimensional manifold $Y$ whose boundary is $\Sigma$. The action of $Z$ satisfies involutory, multiplicative, and associative axioms. In addition, $Z(\emptyset) = C$ for the empty $d$-manifold.

"The *physical* interpretation of [these] axioms," Atiyah goes on to write, is this: "for a closed $(d + 1)$ manifold $Y$, the invariant $Z(Y)$ is the partition function given by some Feynman integral." It is clear that $Z(Y)$ is an invariant assigning a complex number to any closed $(d + 1)$ dimensional manifold $Y$ (in virtue of the fact that the boundary is empty) and clear thus that $Z$ and $P$ coincide.

This definition has two virtues. It draws a relatively clear distinction between parts of a complex theory; and it provides for an interpretation of the mathematical applications along the lines suggested by Section §1. It is less clear, however, in what sense $P$ is a *physical* interpretation of $Z$, the distinction between $Z$ and $P$ appearing to an outsider (this one, at any rate) to have nothing whatsoever to do with any relevant sense of the physical, however loose. The distinction between the mathematical and the physical would seem no longer to reflect any intrinsic features either of mathematical or physical objects, things or processes, with *physical* a name given simply to the portions of a theory that are confused, poorly developed, largely intuitive, or simply a conceptual mess.

In quite another sense, the distinction between the mathematical and the physical is sometimes taken as a reflection of the fact that mathematical objects are quite typically general, and physical objects, specific or particular. The theory of differential equations provides an example. The study of specific systems of equations may conveniently be denoted a part of theoretical physics; the study of generic differential equations, a part of mathematics. But plainly there is no ontological difference between these subjects, only a difference in the character of certain mathematical structures. And this, too, is a distinction internal to mathematics.

The project of determining a clear sense in which mathematics has an application beyond itself remains where it was, which is to say, unsatisfied.

## §11

"To present a theory is to specify a family of structures," Bas van Fraassen has written,

> its models; and secondly, to specify certain parts of those models (the empirical substructures) as candidates for the direct representation of observable phenomena. The structures which can be described in experimental and measurement reports we can call appearances: the theory is empirically adequate if it has some model such that all appearances are isomorphic to empirical substructures of that model.[25]

Some definitions. A language **L** is a structure whose syntax has been suitably regimented and articulated—variables, constants, predicate symbols, quantifiers, marks of punctuation. A language in *standard formulation* has the right kind of regimentation. A model $\mathbf{M} = \langle D, f \rangle$ is an ordered pair consisting of a non-empty domain D and a function $f$. It is $f$ that makes for an *interpretation* of the symbols of L in **M**. Predicate symbols are mapped to subsets of D; relation symbols to relations of corresponding rank on D. The general relationship of language and the world model theory expresses by means of the concept of satisfaction. This is a relationship that is purely abstract, perfectly austere. Formulas in L are satisfied in **M**, or not; sentences of L are true or false in **M**. Languages neither represent nor resemble their models. The scheme is simple.

There are no surprises.

Let **T** be a theory as logicians conceive things, a consistent set of sentences; and a theory furthermore that expresses some standard (purely) mathematical theory—the theory of linear differential operators, say. If **T** has empirical content, it must have empirical consequences—$\varphi$, for example:

**11.1)** $T \models \varphi$.

But equally, if **T** has empirical content, some set of sentences **T(E)** $\subseteq$ **T**, must express its empirical *assumptions*. Otherwise, **11.1** would be inscrutable. Subtract **T(E)** from **T**. The sentences **T(M)** $\subseteq$ **T** that remain are purely mathematical.

Plainly

**11.2)** $T = T(M) \cup T(E)$.

And plainly again

**11.3)** $T(M) \cup T(E) \models \varphi$,

whence by the deduction theorem,

**11.4)** $T(M) \models T(E) \rightarrow \varphi$.

The set of sentences $\Theta = T(E) \rightarrow \varphi$ constitutes the *empirical hull* of **T**.

If model theory is the framework, the concept of a mathematical application resolves itself into a relationship between theory and model and so involve a special case of satisfaction. **T(M)** as a whole is satisfied in a set theoretic structure **M**; $\Theta$, presumably, in another structure **N**, its domain consisting of physical objects or bodies. But given **11.4**, $\Theta$ is also satisfied in *any* model **N** satisfying **T(M)**. **N** is a model with a kind of hidden hum of real life arising from the elements in its domain. An application of mathematics, if it takes place at all, must take place in the connection between **M** and **N**. An explanation of this connection must involve two separate clauses, as in **11.5**, which serves to give creditable sense to the notion of a mathematical application in the context of model theory:

**11.5)** **T(M)** applies to **N** if i) **T(M)** is satisfied in **N**; and, ii) **N** is a sub-model of **M**.

This leaves one relationship undefined. Sub-models, like sub-groups, are not simply substructures of a given structure. If **N** is a sub-model of

M, the domain D' of N must be included in the domain D of M; but in addition:

    i) Every relation R' on D' must be the restriction to D' of the corresponding R on D; and

    ii) ditto for functions; and moreover,

    iii) every constant in D' must be the corresponding constant in D.

This definition has an undeniable appeal. The mathematical applications find their place within the antecedently understood relationship between theories and their models. This does not put mathematics directly in touch with the world, but with its proxies instead. The parts of the definition cohere, one with the other. It is obviously necessary that T(M) has empirical consequences. Otherwise there would be no reason to talk of applications whatsoever. It is necessary, too, that N be a sub-model of M; otherwise the connection between what a theory implies and the structures in which it holds would be broken. Finally, it is necessary that T(M) be satisfied in N as well as M; otherwise what sense might be given to the notion that T(M) *applies* to any empirical substructure of M at all? Those conditions having been met, a clear sense has been given to the concept of a mathematical application.

This is somewhat too optimistic. M, recall, is a mathematical model, and N a model that is not mathematical: the elements in its domain are physical objects, in some sense of the physical, however loose. The assumption throughout is that in knowing what it is for a mathematical theory to be satisfied in M, the logician knows what it is for that same theory to be satisfied in N. In a purely formal sense, it must, that assumption, be true; the *definition* of satisfaction remains constant in both cases. What remains troubling is the question whether the conditions of the definition are ever met. The definition of satisfaction, recall, proceeds by accretion. A sentence S is satisfied in N under a given interpretation of its predicate symbols S[F,G, ... ,H]. The interpretation comes first. In the case of a pure first-order language, it is the predicate symbols that carry all the mathematical content.

Were it antecedently clear that **S[F,G,... ,H]** admits of physical interpretations, why did we ever argue?

# §12

Under one circumstance, the question whether a mathematical theory is satisfied in a physical model may be settled in one full sweep. A theory **T** satisfied in *every* one of the sub-models of a model **M** of **T** is satisfied in particular in the empirical sub-models of **M** as well. It must be. Demonstrate that **T** *is* satisfied in every one of its sub-models and what remains, if **11.5** is to be justified, is the correlative demonstration that **T** is satisfied in a model containing empirical structures as sub-models. What gives pause is preservation itself.

The relevant definition:

**12.1)** A theory *T* is preserved under sub-models if and only if *T* is satisfied in any sub-model of a model of *T*.

In *any* sub-model, note. Preservation under sub-models is by no means a trivial property.

Given **12.1**, it is obvious that preservation hinges on a sentence's quantifiers. A sentence is in prenex form if its quantifiers are in front of the matrix of the sentence; and universal if it is in prenex form and those quantifiers are universal. It is evident that every sentence may be put into prenex form.

Herewith a first theorem on preservation:

**12.2)** If $\varphi$ is universal and N is a sub-model of M, then if $\varphi$ is satisfied in *M* it is satisfied as well in *N*.

The proof is trivial.

**12.1** takes a more general form in **12.3**:

**12.3)** A sentence $\varphi$ is preserved under sub-models if and only if $\varphi$ is logically equivalent to a universal sentence.

Again, the proof is trivial.

**12.3** is, in fact, a corollary to a still stronger preservation theorem, the only one of consequence. A theory **T** has a set of axioms A just in case A and **T** have the same consequences; those axioms are *universal* if each axiom is in prenex normal form, with only universal quantifiers figuring. A theory derivable from universal axioms is itself universal. Let **T**, as before, be a theory. What follows is the Los-Tarski theorem:

**12.4)** **T** is preserved under sub-models if and only if **T** is universal.

**12.4**, together with **11.1**, mark a set of natural boundaries of sorts. **11.1** indicates *what* is involved in the concept of an application; **12.4** specifies *which* theories are apt to have any applications at all. The yield is discouraging. Group theory is *not* preserved under sub-models; *neither* is the theory of commutative rings, *nor* Peano arithmetic, *nor* Zermelo Fraenkel set theory, *nor* the theory of algebraically closed fields, *nor* almost anything else of much interest.

## §13

When left to his own devices, the mathematician, no less than anyone else, is apt to describe things in terms of a natural primitive vocabulary. Things are here or there, light or dark, good or bad. The application of mathematics to the world beyond involves a professional assumption. And one that is often frustrated. Sheep, it is worthwhile to recall, may be collected and then counted; not so plasma. Set theory, I suppose, marks the point at which a superstitious belief in the palpability of things gives way. Thereafter, the dominoes fall rapidly.

If some areas of experience seem at first to be resistant to mathematics, there is yet a doubled sense in which mathematics is inexpungable, a feature of every intellectual task. The first of these senses arises as the result of a piece of metaphysical jujitsu. The idea that there is some arena in which things and their properties may be directly apprehended is incoherent. Any specification of the relevant arena must be by means of some theory or other; there is no describing without descriptions. But to specify a theory is to specify its models. And so mathematics buoyantly enters into areas from which it might have been excluded, if only for purposes of *organization*.

Mathematics makes its appearance in another more straightforward sense. Every intellectual activity involves a certain set of basic and ineliminable operations of which counting, sorting, and classification are the most obvious. These operations may have little by way of rich mathematical content, but at first cut they appear to be amenable to formal description. It is here that the empirical substructures that van Fraassen evokes come into play. Mention thus of *empirical substructures* is a mouthful; let me call them *primitive models*, instead, with *primitive* serving to emphasize their relative position on the bottom of the scheme of things, and *models* reestablishing a connection to model theory itself. The primitive models are thus a mathematical presence in virtually every discipline, both in virtue of their content—they are models, after all; *and* in virtue of their form—they deal with basic mathematical operations.

Dana Scott and Patrick Suppes envisage the primitive models as doubly finite: their domain is finite; so are all model-definable relations. Those relations are, moreover, *qualitative* in the sense that they answer to a series of *yes* or *no* questions asked of each object in the domain of definition.[26] This definition reflects the fact that in the end every chain of assertion, judgment, and justification ends in a qualitative declaration. There it is: the blotting paper is red, or it is not; the balance beam is to the right, or it is not; the rabbit is alive, or it is dead. But now a second step. A physical object is any object of experience; and objects of experience are those describable by primitive *theories*. Primitive theories are satisfied in primitive models.

The salient feature of a primitive model is a twofold renunciation: only finitely many objects are considered; and each object is considered only in the light of a predicate that answers to a simple *yes* or *no*. Such are the primitive properties $F_1, \dots, F_n$. A primitive model may also be described as any collection C of primitive properties, together with their union, intersection and complement.

There is yet another way of characterizing primitive theories and their models. A *Boolean-valued* function $f$ is one whose domain is the collection of all $n$-tuples of 0 and 1, and whose range is $\{0, 1\}$. Such functions are of use in switching and automata theory. Their structure makes them valuable as instruments by which qualitative judgments are made and then recorded. The range of a Boolean-valued function corresponds to the simple *yes* or *no* (true or false) that are features of the primitive models; but equally, the domain corresponds either to qualitative properties or to collocations of such properties. A primitive theory, on this view, is identified with a series of Boolean equations; a primitive structure, with a Boolean algebra.

Set theory provides still another characterization of the primitive models, this time via the concept of a generic set. The generic sets are those that have only the members they are forced to have and no others. Forcing is atomistic and finite, the thing being done piecemeal. Thus suppose that $L$ is a first-order language, with finitely many predicates but indefinitely many constants. By the extension $L^*(S_1, ..., S_n)$ of $L$, I mean the language obtained by adjoining the predicate symbols $S_n$ to $L$. A *basic sentence* of $L^*$ has the form $k \in S_n$ or $k \notin S_n$. A finite and consistent set of basic sentences $\xi$ constitutes a condition. A sequence of conditions $\xi_1, ... , \xi_n$ is *complete* if and only if its union is consistent, and, moreover, for any $k$ and $n$, there exists an n such that $k \in S_n$ or $k \notin S_n$ belongs to $\xi_n$. A complete sequence of conditions determines an associated sequence of sets $S_1, ... S_n$:

$$k \in S_n \Leftrightarrow (\exists m)(k \in S_n \text{ belongs to } \xi_n).$$

The path from conditions to sets runs backwards as well as forwards.

Sets have been specified, and sequences of sets; conditions, and sequences of conditions. The model structure of $L^*$ is just the model $M^* = \,<D, f>$, where $f$ maps $S_1, ..., S_n$ onto $S_1, ..., S_n$. There is a straightforward interpretation of the symbolic apparatus. The conditions thus correspond to those *yes* or *no* decisions that Scott and Suppes cite; that they are specified entirely in terms of some individual or other belonging to a set is evidence of the primacy of set formation in the scheme of things.

The specification of sets by means of their associated conditions is a matter akin to enumeration. A given set S is *generic* if in addition to the objects it has as a result of enumeration, it has only those objects as members it is forced to have. Forcing is thus a relationship between finite conditions and arbitrary sentences of $L$, the sentences in turn determining what is truly in various sets. The definition proceeds by induction on the length of symbols. What it says is less important than what it implies. Every sentence about a generic set is decidable by finitely many sentences of the form $k \in S_n$ or $k \notin S_n$. Finitely many, note, and of an *atomic* form.

Whatever the definition of primitivity, theories satisfied in primitive models admit of *essential application*. This is a definition:

**13.1)** $T$ *applies essentially* to M if and only if M is primitive and $T$ is satisfied in M.

Such theories apply directly to the world in the sense that no other theories apply more directly. Counting prevails, but only up to a finite point; the same for measurement. The operation of assigning things to sets is suitably represented; and in this sense one has an explanation of sorts for the universal feeling that numbers may be directly applied to things in a way that is not possible for groups or Witten functors. It is possible that the operation of assigning things to sets is the quintessential application of mathematics, the point that dizzyingly spins off all other points. But even if assigning things to sets is somehow primitive, the models that result are themselves abstract and mathematical.

The concept of a primitive model does not itself belong to model theory. The primitive models have been specified with purposes other than mathematics in mind. Nor is it, that concept, precisely defined, if only because so many slightly different structures present themselves as primitive. Nonetheless, the primitive models share at least one precisely defined model-theoretic property. A number of definitions must now be introduced:

Let $L$ be a countable language and consider a theory $T$:

**13.2)** A formula $\varphi(x_1, x_2, \dots, x_n)$ is *complete* in $T$ if and only if for every other formula $\chi(x_1, x_2, \dots, x_n)$, either $\varphi \supset \chi$ or $\varphi \supset \sim\chi$ holds in $T$; and

**13.3)** A formula $\theta(x_1, x_2, \dots, x_n)$ is *completable* in $T$ if and only if there is a complete formula $\varphi(x_1, x_2, \dots, x_n)$ such that $\varphi \supset \theta$ holds in $T$.

The definitions of complete and completable formulas give rise, in turn, to the definitions of atomic theories and their models:

**13.4)** A theory $T$ is *atomic* if and only if every formula of $L$ consistent with $T$ is completable in $T$; and

**13.5)** A model M is *atomic* if and only if every relation in its domain satisfies a complete formula in the theory $T$ of M.

It follows from these definitions that every finite model is atomic; it follows also that every model whose individuals are constant is again atomic. Thus a first connection between empirical substructures and model theory emerges as a trivial affirmation:

**13.6)** Every primitive model is atomic.

The proof is a matter of checking the various definitions of the primitive models, wherever they are clear enough to be checked.

The real interest of atomic models, however, lies elsewhere. The relationship of a model to its sub-models is fairly loose; not so the relationship of a model to its *elementary* sub-models. Consider two models, N and M, with domains $D'$ and $D$. A first-order language $L$ is presumed throughout:

**13.7)** The mapping $f: D' \to D$ is an *elementary embedding* of N into M if and only if for any formula $\varphi(x_1, x_2, \dots, x_n)$ of $L$ and $n$-tuples $a_1, \dots, a_n$ in $D'$, $\varphi[a_1, \dots, a_n]$ holds in N if and only if $\varphi[fa_1, \dots, fa_n]$ holds in M.

Given **13.7**, it follows that the target of $f$ in M is a sub-model of M; the elementary sub-models are simply those that arise as the result of elementary embeddings. From the perspective of first-order logic, elementary sub-models and the models in which they are embedded are indiscernible: no first-order property distinguishes between them. The models are *elementarily equivalent*.

Another definition, the last. Let N be a model and $T(N)$ its theory:

**13.8)** N is a *prime model* if and only if N is elementarily embedded in every model of $T(N)$.

**13.6** establishes trivially that every primitive model is atomic. It is also trivially true that primitive models are countable. But what now follows is a theorem of model theory:

**13.9)** If N is a countable atomic model, then N is a prime model.

Assume that N is a countable atomic model; $T(N)$ is its theory. Say that $A = \{a_0, a_1, \ldots, a_n\}$ constitutes a well-ordering of the elements in the domain of N. Assume that M is any model of $T(N)$. Suppose that $F$ is a complete formula satisfied by $a_0$. It follows that $(\exists x)F$ follows from $T(N)$; it follows again that there is a $b_0$ among the well-ordered elements B of M that satisfies $F$. Continue in this manner, exhausting the elements in $A$. By **13.7**, going from $A$ to $B$ defines an elementary embedding of N into M. The conclusion follows from **13.8**.

**13.6** establishes that the primitive models are among the atomic models; but given the very notion of a primitive model, it is obvious that any primitive model must be countable. It thus follows from **13.9** that

**13.10**) Every primitive model is prime.

In specifying a relationship between the primitive and the prime models, **13.10** draws a connection between concepts arising in the philosophy of science and concepts that are model-theoretic. There is a doubled sense in which **13.10** is especially welcome. It establishes the fact that the primitive models are somehow at the bottom of things, in virtue of **13.8** the *smallest* models available. And it provides a *necessary* condition for a theory to have empirical content. Recall van Fraassen's definition: "A theory is empirically adequate if it has some model such that all appearances are isomorphic to empirical substructures of that model." A mathematical theory $T$ has empirical content just in case $T$ has a prime model.

Such is the good news. The bad news follows. **13.10** does little—it does *nothing*—to explain the relationship, if any, between those mathematical theories that are not primitive and those that are. As their name suggests, the primitive models are pretty primitive. The renunciations that go into their definition are considerable. There is thus no expectation that any mathematical theory beyond the most meager will be definable in terms of a primitive theory. Let us go, then, to the next best thing. Assume that most mathematical theories are satisfied in models with primitive sub-models. **13.6** might then suggest that such theories apply to elements in a primitive sub-model if they are satisfied in a primitive sub-model. But those mathematical theories that are not preserved under sub-models generally will not be preserved generally under primitive sub-models either. The primitive theories and their models are simply too primitive.

There is no next best thing.

## §14   Conclusion

The argument that mathematics has *no* application beyond itself satisfies an aesthetic need: it reveals mathematics to be like the other sciences and so preserves a sense of the unity of intellectual inquiry. Like any argument propelled by the desire to keep the loose ends out of sight, this one is vulnerable to what analysts grimly call the return of the repressed. Mathematics may well be akin to zoology; yet the laws of physics, it is necessary to acknowledge, mention groups and not elephants. And mathematical theories in physics are strikingly successful. Alone among the sciences, they permit an uncanny epistemological coordination of the past, the present, and the future. If this is not evidence that in some large, some irrefragable sense, mathematical theories apply to the real world, it is difficult to know what better evidence there could be.

That mathematical objects exist is hardly in doubt. What else could be meant by saying that there exists a natural number between three and five? Where they exist is another matter. The mathematical Platonist is often said to assert that mathematical objects exist in a realm beyond space and time, but since this assertion involves a relationship that is itself both spatial and temporal, it is very hard to see how it could be made coherently.

The idea that mathematical objects are the free creations of the human mind, as Einstein put it, is hardly an improvement. If the numbers are creations of the human mind, then it follows that without human minds, there are no numbers. In that case, what of the assertion that there is a natural number between three and five? It is true now; but at some time before the appearance of human beings on the earth, it must have been false. The proposition that there exists a natural number between three and five cannot be both true and false, and so it must be essentially indexical, its truth value changing over time. That Napoleon is alive is accordingly true during his life and false before and afterwards. But if the proposition that there exists a natural number between three and five is false at some time in the past, the laws of physics must have been false as well, since the laws of physics appeal directly to the properties of the natural numbers. If the laws of physics were once false, of what use is any physical retrodiction—any claim at all about the distant past?

Perhaps then mathematical assertions are such that once true, they are always true? This is a strong claim. On the usual interpretation of modal logics, it means that if $P$ is true, then it is true in every possible world. Possible worlds would seem no less Platonic than the least Platonic of mathematical objects, so the improvement that they confer is not very obvious.

Various accounts of mathematical truth and mathematical knowledge are in conflict. The truths of mathematics make reference to a domain of abstract objects; they are not within space and they are timeless. Contemporary theories of knowledge affirm that human agents can come to know what they know only as the result of a causal flick from the real world. It is empirical knowledge that is causally evoked. Objects that are beyond space and time can have no causal powers.

To the extent that mathematical physics is mathematical, it represents a form of knowledge that is not causally evoked. To the extent that mathematical physics is not causally evoked, it represents a form of knowledge that is not empirical. To the extent that mathematical physics represents a form of knowledge that is not empirical, it follows that the ultimate objects of experience are not physical either.

What, then, are they? As a physical subject matures, its ontology becomes progressively more mathematical, with the real world fading to an insubstantial point, a colored speck, and ultimately disappearing altogether. The objects that provoke a theory are replaced by the enduring objects that sustain the theory. Pedagogy recapitulates ontology. The objects treated *in* classical mechanics, to take a well-known example, are created *by* classical mechanics. Unlike the objects studied in biology, *they* have no antecedent conceptual existence. In V. I. Arnold's elegant tract, for example, a mechanical system of $n$ points moving in three-dimensional Euclidean space is defined as a collection of $n$ world lines. The world lines constitute a collection of differentiable mappings. Newton's law of motion is expressed as the differential equation $x'' = F(x, x', t)$.[27]

Nothing more.

Mathematics is not applied to the physical world because it is not applied to anything beyond itself. This must mean that as it is studied, the physical world becomes mathematical.

# VI. TITANS

# 25. Isaac Newton

It is the aim of physics to limn the ultimate structure of reality by means of laws that are at once general and simple. This lapidary formulation raises precisely seven problems: What is the meaning of "limn," "ultimate," "structure," "reality," "laws," "general," and "simple"? I have no idea; and neither, I suspect, do the physicists.

## Newton's Version

The appeal of the Newtonian vision is much a matter, I think, of a glimpse glanced of a world that is clear in all of its aspects, unchanging, measured, determined, and regular. My own life may tremble precariously on the very cockroach cusp of chaos, but the Newtonian universe is bisected by "Absolute, True and Mathematical Time, which of itself, and from its own nature, flows equably without regard to anything external; [and] Absolute Space, which in its own nature, without regard to anything external, remains always similar and immovable." This is a formulation in which a man may find comfort without worrying overmuch about its coherence. Newtonian *mechanics*, as it has turned out, is incorrect in that bizarre limit in which objects contract, mass expands, and time itself slows and then stops. This circumstance is relevant only to those who demand of a theory that it be true. And besides, beyond Newton's version of mechanics, there is Newton's vision.

From one perspective, Newtonian mechanics is a theory of motion; from yet another, an elaborate exercise in the classification of certain geometrical figures. Under the spell of either incarnation, ordinary objects shed with fascinating speed their habitual properties of color, density, and texture: A billiard ball or a planet becomes a *point-mass*, the whole of its now perilous identity concentrated at its center. The Newtonian systems thus occupy rather a sparse conceptual stage. They tend to appeal to the man with a taste for desert landscapes.

The computer, I have observed, marks time in integral instants. Between the beats of its mechanical heart or clock, the machine has no life and slips into a temporary void. Newtonian time, by way of contrast, is continuous. Between any two temporal points, there is a third. These touchstones (or tombstones) go on forever. This image of Newtonian time may well suggest misleadingly that on this scheme time has a definite direction. Not so. The equations of motion that figure in Newtonian mechanics are *time-independent* (the physicist's term) or *autonomous* (the mathematician's); what has come and what is to come coincide, providing the physicist with his first, and only, view of the natural world under an aspect of eternity.

Wherever they are and whatever they are doing, the elements of a Newtonian system occupy some quite definite position and so establish themselves as a geometric *configuration*. The set of such configurations constitutes an abstract space of possibilities thickly enveloping any actual configuration of Newtonian elements in the here and now. Observing a particular Newtonian system—the solar system, for example—the physicist intervenes in its workings (intellectually if not literally) only to fix formally the system's *initial configuration*; he then aims to determine completely the future position of the particles and describe the trajectory of their motion. Doing the second, he has done the first.

On the level of metaphysics, it is apparent that things change only because something gets them to change. It is in the nature of physical curiosity to wish to make this declaration precise. Our own experiences on the surface of the Earth suggest that everything that moves is moved by something, and requires thus a continual infusion of force if it is to remain in motion. Such was Aristotle's dictum—the doctrine of impetus. To the extent that it is true, it is plainly true *locally*, where baseballs and ballerinas rise in a graceful arc, their velocity diminishing in proportion to the force with which they were originally impressed, and the resistance that they encounter—the friction of the air, say.

There is something impossibly tender in the long tradition of Aristotelian mechanics. Absolutely nothing to which one can sensuously point indicates *obviously* that the Aristotelian view of things is incorrect. The discovery that it is occurs in the history of thought as a rude shock, and marks the point at which Nature and trust in the obvious collide, with trust in the obvious very much the bruised loser. The relationship between force and motion that Aristotle advanced, Newton rejected. It is only a change in velocity—*acceleration*—that is the result of an application of force—$F = ma$, to put the whole thing into four famous symbols. Where no force is needed, none need be postulated. In the absence of friction, an object moving in a straight line will continue to move in a straight line forever. In the absence of force, an object at rest—an undergraduate, say—will remain at rest forever.

§

As the physicist sees things, Newtonian mechanics is a branch of physics, and limited thus to the one (and only) physical world. The physicist is an intellectual Puritan in the narrowness of his affections. Yet there are as many mathematical worlds as there are consistent mathematical theories. The mathematician is an intellectual Mormon in the voluptuousness of *his* attachments and is forever appearing in public accompanied by ten lavishly complaining wives. Strangely enough, mathematics and physics are mutually absorbed; that these two separate disciplines should have anything to do with one another is a very great mystery.

Like any other science, mathematics begins with certain concepts too rich to admit of definition—the notion of a *function*, for example, something that serves to pair one group of objects with certain Significant Others. A function is thus a relationship between two or more mathematical objects. These need not be numbers. Unfortunately, the notion of a relationship is no clearer than the notion of a function.

Relationships and functions may both be defined in the terms of set theory, a subject, as I have said, astonishingly rich in paradox and little else, but there the chain of definitions simply stops.

The simplest of all functions is also the most obvious and acts to pair a number with itself. Under its influence, the number 1 is mapped to the number 1, 2 to 2, and 999 to 999. The function acts as a mirror. In squaring a number, to consider another function, I pair a number to itself, multiplied just once by itself: 1 to 1, 2 to 4, 3 to 9. To throw the relationship into relief, mathematicians write $f(1) = 1$, $f(2) = 4$, and $f(3) = 9$. Here $f$ is presumed to operate on 1, then 2, and then 3, and may be imagined as a kind of magician who, after draping parentheses around a given number, diverts the eye just slightly to the right, where a brand new number miraculously appears.

The process involved in pairing a number to itself may be expressed in its full and virile generality by means of algebraic variables—letters such as $x$, $y$, or $w$, which stand indiscriminately for any natural number: $f(x) = x^2$, instead of $f(1) = 1$ or $f(2) = 4$. This is a handy form of shorthand, in which an operation of infinite scope is given by what amounts to a simple military command: take any natural number, Buster, and square it.

§

The drama of Newtonian mechanics plays within a space of three dimensions—an old standby from daily life—up and down, left and right, in and out; but however much we may wonder why our physical *experience* inevitably takes place in precisely three dimensions, there is nothing in the mechanics of the matter that makes three an especially important number. Within special and general relativity, the three dimensions of space jostle companionably for attention in the company of a fourth dimension marking time.

A straight line constitutes a space of but one mathematical dimension. Its analysis is complete when its origin has been specified, together with a unit of measurement. Imagine now two infinitely long straight lines crossed at the perpendicular. These are the axes of a two-dimensional *Cartesian coordinate system*. Each of those straight lines is numerically sub-divided. At their point of intersection, both are set to 0. The fact that these lines intersect makes it possible to represent negative as well as positive numbers in space. But a Cartesian coordinate system also depicts the plane, which is purely a geometric object, and so forges a connection between two parts of the mathematical experience—geometry and algebra. Each point in the plane, for example, may be uniquely identified by a pair of numbers—its coordinate—by the simple device of counting along each axis. With the addition of a third coordinate axis, a mathematical representation of space itself emerges. The specification of a point now requires three separate numbers. In elementary physics textbooks, triplets give way to *vectors*, a class of curiously rootless geometrical objects; but vectors and numbers admit of a mathematical merger or marriage. A two-dimensional vector is nothing more than a pair of numbers; a three-dimensional vector is a triplet; and an *n*-dimensional vector, n numbers frozen in a particular order.

Very simple mathematical ideas lie at the secret silent heart of physics. I shall elaborate. Newton thought to compare a particle to a mathematical point, and its trajectory—where it goes, where it has been—to a curve throughout the whole of space. This was an act of double abstraction and a queer initial point for a theory of motion: points have no width or mass, and curves do not move. In any event, suppose that a particle passes through the origin of a two-dimensional Cartesian coordinate system bound for parts unknown. To the right is an axis marking time; above, an axis marking distance (but not direction). The speed of the particle is constant. The physicist achieves a *partial* physical description of the particle's behavior if he is able to say for a given time how far the particle has travelled from its origin; a *complete* description, if he is able to describe distance in terms of time. A relationship of this sort— distance against time—requires a function for its expression. And functions are *mathematical* objects.

The simplest (although not the most common) curve in space is a straight line. Strictly speaking, lines and curves are geometric objects and belong to a primitive intellectual kingdom. In analytic geometry, a subject no doubt remembered with a moan of pure misery by almost anyone past high school age, a lucid and invigorating connection is created between certain algebraic equations (the moan deepens) and a class of curves (the moan darkens), the effect as striking as that produced by an illustrated medieval manuscript. Pairs of distinct points in the plane (*four* numbers are required) determine a single straight line; there the thing hangs, dutifully extending itself forever in two directions. Straight lines may also be described algebraically by an equation: $y = mx + b$, for example, where $b$ is the point at which the line crosses the $y$-axis, and $m$ represents its *slope*, or angle of inclination. Four numbers are again required to express this equation; three to solve it.

In two dimensions, straight lines may ascend or descend, move vertically upward like a rocket, or remain horizontal like a thin stream crossing the plane from one remumbled infinity to another. The slope of a straight line, I have said, corresponds to its angle of inclination—the ratio of differences between successive values of $y$ and $x$. A line pitched at a forty-five degree angle moves upward by one unit for every unit that it moves to the right. Its slope is 1. That plane, with its ascending straight line and crossed and martyred coordinate axes, the reader must now hold in suspension and flash-freeze as an image. A cyclist, imagine, dressed, perhaps, in those absurd and iridescent tights much in favor in California, is bent on moving from here to there. His motion I reflect by means of the mathematical mirror of a coordinate system—the very one the reader is now instructed to fetch from memory. Movement along the $x$-axis, as before, is movement in time; the $y$-axis measures distance. Units are fixed first in minutes and then in miles; the cyclist's behavior I reduce to, and then represent by, the trajectory of a curve. The routine ratio of distance over time now acquires a novel incarnation as *velocity*, a concept with no cognate in mathematics. Two minutes having elapsed, the cyclist has covered two miles. His rate of speed is precisely one mile a minute.

Only three physical concepts have made an appearance in this discussion: time, distance, and speed. How far a particle has gone is a matter of how fast it has been going, and how long it has been going fast. How fast a particle is going is a matter of how far it has gone, and how long it has taken to go far. Time and either speed or distance suffice to make the real world rise.

## The Vagaries of Curvature

Newton's law of motion associates force, mass, and acceleration. An object in uniform and rectilinear motion experiences no acceleration and moves across the coordinate plane (one axis representing time, the other distance) with smooth and untroubled confidence. It is thus, no doubt, that objects move in Hell. In determining the slope of a straight line, mathematicians take the tritest of ratios; because the line is straight, it does not matter *where* the slope is measured. Throughout the whole of space it stays stubbornly the same. Now, a straight line is the shortest distance between two points, but not the most common, as airline pilots and politicians are both aware. And, unlike a straight line, a curve is a creature of change. The equation $f(x) = x^2$ describes a parabola in the plane. At 1, $f(x)$ is 1, at 2, 4, and at 3, 9. The obvious procedure for fixing the slope of *this* line by taking ratios fails. Different points make for different ratios. But then again those physical objects that move without changing their motion are singular. Rising from the couch with a yawn and a stretch, *I* pass from a state of rest to a state of vigorous and untroubled activity. Acceleration. Reconsidering the whole business, I sink back onto the couch with a sigh and a moan. Deceleration. By means of a single superbly controlled effort (really rather like a panther, I have been told) I go quite beyond uniform and rectilinear motion. The equation that might describe what I have done yet again depicts a curve in space.

To the physicist, velocity appears as a *physical* concept arising spontaneously from purely a *mathematical* circle of definitions. So far, the mathematician has made himself clear only in the case of the straight line. The case of the common curve remains beyond a definition of velocity framed in terms of straight lines and their slopes. Still, some content, the physicist might urge, may yet be given to the *general* concept of velocity by averaging speeds. The average velocity of a particle that has covered sixty miles in sixty minutes is just sixty miles an hour. This has all the appearances of a sane and sensible solution to the problem of acceleration. A particle in motion may, however, begin things in a state of rest and yet cover sixty miles in sixty minutes by means of a late, lavish burst of speed. On distributing the particle's average velocity to each of those sixty minutes, the physicist is apt to conclude that at any particular moment the damn thing is either moving faster or slower than it really is. This is an unhelpful position to have reached. What the physicist instead requires is the *instantaneous* velocity of a particle—some measure of its speed *now*.

In considering precisely this problem, Leibnitz and Newton thought to rescue the idea of a ratio. Not worried overmuch by logical scruples, they accepted without qualm the thesis that some numbers might be less than any other number and yet greater than zero. Such are the *infinitesimals*. The ratio of infinitesimal change in distance over infinitesimal change in time they reckoned the velocity of an object at a point. It was by this act that a suspension of mathematical disbelief was achieved. The great analytic mathematicians of the nineteenth century managed to eliminate this appeal to the logically unwholesome by means of the concept of a *limit*—one of the great, fabulous, fragrant flowers in the history of thought. The fractions $\frac{1}{1}$, $\frac{1}{2}$, $\frac{1}{3}$ when continued in the obvious way (the denominator getting larger), constitute an infinitely extended mathematical series. Somewhere beyond any of the fractions themselves, solitary, singular, seductive, is the limit toward which those numbers are patiently plodding—0, in the present case. In the calculus, this elegant but general concept is specialized. At 2, $f(2)$ is 4. I retain the example of the parabola, but the action is now localized at a point. Suppose that $h$ represents a small increment to 2 – 2 and a tad, so to speak. *This* increment is real and shares no disturbing kinship to those infinitesimals employed by Leibnitz and Newton. The ratio between $f(2 + h) - f(2)$ and $h$ is a measure of the small degree in which changes in one direction along the axis of a curve are balanced by small changes in the other. Assigning this ratio to the curve as its slope, the mathematician remains in the domain of average quantities. Yet what happens when *h* *contracts?* Recall that *f* maps each number to its square. Hence $f(2 + h) - f(2) = (2 + h)(2 + h) - 4 = 4 + 4h + h^2 - 4$. When divided by *h*, this comes to $4 + h$. As *h* becomes smaller, the sum tends simply to 4. Tendings and tendencies suggest the existence of a limit. Precisely. A limit so defined is a *derivative* and functions purely as a mathematical measure of curvature; when reinterpreted physically, derivatives serve as stand-ins for the concept of instantaneous speed.

The calculation that I have just carried out may be extended so that each point on the curve is assigned a slope. In general terms, with variables standing in for particular numbers, *f* is $f(x) = x^2$. Consider now the behavior of $f(x + h) - f(x)$. This amounts to $(x + h)(x + h) = x^2 + 2xh + h^2 - x^2$. First and last figures in the numerator cancel; divide by *h*, and the result is $2x + h$. As *h* itself heads for home at zero, $2x + h$ reduces to the limit $2x$. The derivative of a function at a point is a number; the derivative of a function as a whole is yet again a *function.*

Just recently, I must mention, simply to complete the story, infinitesimals have made a reappearance, purged by contemporary logicians of their impurities. It is now possible to picture the slope of a curve in the old-fashioned way, a strange development in the history of mathematics in which a queer path is traced from the absurd to the austere and back again to the absurd.

## Celestial Dynamics

Force impresses itself on matter by acceleration. This marks the difference from moment to moment in an object's speed. The velocity of any object is its change in position against its change in time. Acceleration is thus the derivative of velocity. Newton's second law of motion states that force is given as the product of mass and acceleration. Knowing the forces that impinge upon an object, the physicist may calculate its acceleration, and with its acceleration, its velocity, and ultimately, by means of integration (another radiant, multi-hued mathematical concept—the cousin and correlate to differentiation), its change in position.

All of this lies somehow on that tentative border between mathematics and physics, and offers a contrast between two sets of concepts: the delicate Greek tangle of the calculus, the simple Roman brutality of Newton's law.

It was Newton's vast ambition to provide a complete analysis of the mechanical behavior of a system of interacting particles. What lent to his work its imperial boldness was his insistence that one and the same analysis might apply indifferently to *any* system of interacting particles. This served to unify in one mathematical figure the behavior of objects in motion on the surface of the Earth and the trajectories of the planets themselves, two systems that no one had thought to bring under the impress of a common mathematical description. In a single stunning year, Newton, working alone, and absent from London because of the plague, described and then solved the equations of motion for an interacting system of mechanical particles, and invented the mathematical tools, chiefly the calculus, that were necessary to articulate the whole of his vision.

By means of the peculiar and unrepeatable alchemy of genius, Newton solved completely the abstract problem of describing the universe as a mechanical system, at least in the sense that he was able to state for an indefinite number of particles or planets the laws governing their evolution. Within celestial dynamics, that most musical of sciences, the solar system is treated as a Newtonian mechanical system. In the century before Newton's birth, Johannes Kepler had observed that the orbit of those planets that he could see described an ellipse. Kepler's law, Newton knew, sufficed to calculate the acceleration with which the planets revolved around the sun, and with their acceleration, their trajectory. It was in the course of carrying out this calculation that Newton was led to the inverse-square law. All objects in the universe attract one another, Newton discovered (or decided—the line between fact and fabrication is very thin in physics), with a force inversely proportional to the square of the distance between them, and proportional to their mass. The universal force of gravitation, although affected *by* distance, acts *at* a distance, through empty space and across the inhospitable regions of the stellar night. The inverse-square law holds not only for the sun and its satellites, but for all material bodies wherever they might be found, in the observable regions of the solar system or in the depth of space, where unimaginable cold predominates. By treating the solar system as if it consisted entirely of two bodies—the sun and a planet—Newton was able to demonstrate that to the extent that these bodies were attracting one another in accordance with the inverse-square law, the planet, in tracing a curve through the sky, would describe an ellipse. He thus returned to precisely the intellectual point from which he had departed. There is a very pleasant symmetry at work in all this, an engaging play between fact and theory, with Kepler and Newton, resurrected now in a timeless world, pointing weakly at each other (crooked fingers, faint smiles), like athletes too pooped to do much more than pant.

Within a Newtonian universe change comes about when particles alter their position in space. It is thus that the chugging steam train moves across the landscape of elementary texts, black smoke billowing. From a more abstract point of view, in which there is less to see and more to appreciate, those changes undergone by a Newtonian system involve a systematic process in which *states* of the system instantaneously give way to other states, as when with a certain pressure of thumb and forefinger, the magician gets the cards of a deck to snap forward in a series of pale, repetitive scraps. Whatever intuitive meaning Newton's laws of motion may have—this usually illustrated by ricocheting bullets or billiard balls—they function mathematically as *mechanical* principles and serve in the history of thought to endow the concept of a machine with almost all of the content it ever acquires.

And here is the point to a puzzle. A mechanical system, we are inclined lightheartedly to say, is one that enjoys an unusual analytic transparency. In large measure, we understand the very concept of determinism—as in *I hadda do it* (usually an assassination or assignation)—only by reference to Newtonian mechanics, and not the other way around. What the system does is determined by what it did. Knowing the latter, we know the former. *Pas tout a fait.* The great majority of differential equations do not lend themselves to analytic solution, and lie there ob-durately on the page, the precise and global nature of their destinies inscrutable. In systems comprising two bodies Newton's equations of motion admit of a complete and closed solution. When three particles interact—only one more particle, after all—the Newtonian system that results cannot be solved. Three is the number of the Trinity, and the number, too, at which the universe ceases to be computable—evidence, if any were needed, that in science, as in the rest of life, one is dealing with a form of irony.

## Newton's Vision

At a point roughly midway between Richard Nixon's first inauguration and his hysterical and hilarious farewell speech, it became fashionable among writers and thinkers (a nice distinction, that) to see any num-ber of ecological or social catastrophes coming in the short term. Jay Forrester and Dennis Meadows, for example, studied models of the world's economies and reported themselves convinced that disaster was only decades away; at the *International Institute for Applied Systems Analysis* in Vienna, where I spent a wonderfully beery three months, great global models depicting the production and consumption of en-ergy were created on the computer, with results that suggested that someone had better do something promptly. During the last years of President Carter's luckless administration, systems analysts construct-ed a global model of the world so bleak in its conclusions as to confirm the president in his conviction of the ubiquity of sin and the inevitabil-ity of its punishment.

It goes without saying, of course, that since time immemorial some-one somewhere has stood solemnly in some marketplace or other, warn-ing the rest of us that we had better get our acts together: don't smoke, don't drink, don't do drugs, don't despair—a quartet of remonstrations obviously impossible simultaneously to satisfy. There is an interesting difference, however, between then and now. The Hebrew Prophets were much concerned with sounding an alarm among the Hebrew people.

What happened to primitives elsewhere was only fitfully (and in the case of Jonah, grudgingly) their concern. This has always struck me as a sensible attitude. The first pictures of the Earth from outer space, however, suggested to almost everyone that as far as life goes we are all in this together. Contemporary concerns are thus *global* in that trite and tenuous way in which it is often said that the world has become a global village (Redwood City and not Florence, unfortunately). No doubt, the idea had occurred before: those pictures gave to the idea a dramatic charge. But the process by which a leading theme is translated into a living theory is one crucially contingent on the availability of certain concepts. The business of seeing the world as a single object, in which a great many diverse things and processes hang together, requires some tolerable notion of a *system*. This idea came late to human consciousness, and expresses, if anything, the world view projected so patiently by Newtonian mechanics—Newton's vision as opposed to Newton's version.

No less than any other theory, Newtonian mechanics achieves its singular force by means of a concentration of its intellectual resources. A natural law—$F = ma$, for example, or the inverse-square law—represents the intense, bittersweet liquor that remains when a concept is steadily distilled. The physical world is in its largest aspect informed by the concept of *force*; the rest of the universe (where life, love, luck, and language hold sway), by the concept of *growth*. Within Newtonian mechanics, force is ultimately explained by the laws of force, but in biology, or the social sciences, growth is tied to no correlative natural laws. The very simplest equations of growth, for example, suggest that at any given time an object, or collection of objects, grows in a way that is proportional to its size. When expressed precisely, models of growth admit of a simple solution: things grow exponentially. This way of looking at things is hardly an improvement over the declaration that often things grow, generally until they stop. The mathematics is ceremonial.

Newton's vision, I have suggested, is much a matter of the projection of Newton's version of mechanics onto an alien screen. What has stayed in focus under this projection is simply the concept of a system.

Is there any reason to suppose that anything else will ever come clear? I am inclined to answer my own question with a laconic *nope*; but making this point plain would require a separate argument, another book, another life.

# 26. The Ghost within the Machine

Alan Turing came to Princeton in 1936 to study with the logician Alonzo Church (the same Church with whom I studied, then slim as a steeple). He was nineteen at the time, as elegant as a slept-in shirt, and a classic British ectomorph—I am judging from his photograph, which depicts him in running shoes and shorts, his head at an awkward angle, *les yeux perdus*.

In his temper, Turing belonged in the company of history's great-hearted cranks. This is an image that expresses his inventiveness, his curiosity, even his perseverance. It fails to do justice to the striking mathematical depth of his thought. Along with a handful of other mathematicians and logicians, Turing worked at the very margins of the mathematical experience, unsure whether he would ultimately fall from that perilous edge and vanish into the void.

To the question of how thought proceeds, the normal response, I think, is by a mixture of insight and intuition—a kind of forward lurch that remains phenomenologically insusceptible to specification. Consciousness records a fast blur of tension and release. It is only when thoughts are expressed in language that an activity of the intellect appears amenable to decomposition—thoughts are then mapped to sentences; sentences to words. What remains is thus a finite alphabet of *symbols*.

A machine, I have argued, is the most general of devices taking inputs to outputs by means of a set of states. The human subject begins with words and ends with words. In going anywhere, he has only words to go on. But in passing from one set of words to another (from hearing what he says to saying what he hears, for example) a human agent must make use of a purely physical object. In carrying out any particular computation, Turing reasoned, the brain occupies a particular neurophysiological configuration or state; these the neurophysiologist identifies with a distribution of neurons or even a collection of chemical pathways. And yet, the human brain is finite. So too, then, are the states that it may occupy.

The identification and analysis of the brain's neurological states Turing quite properly thought a task for the biologist. The mathematician need only attend to their *existence*. By means of a great dreamy imaginative leap Turing merged these observations with that definition and concluded that thinking is purely a mechanical process. Having reached this high ground, he wondered whether the machine that instantiated an act of thought need be the human body. This very natural and dramatic question suggested to Turing the idea of a *Turing Machine*.

A Turing machine is designed to manipulate the elements of a set of symbols—words, say, or numbers, or letters. These constitute the machine's alphabet; they are displayed for the machine's examination on a tape divided into squares. The tape stretches infinitely in both directions. Above the tape is a reading head that serves to scan the squares. It may itself move one step to the right, or one step to the left. It may not move at all—an ancient, sphynx-like eye. Symbols from the alphabet the reading head inscribes on the tape, one symbol to a square; those symbols it no longer requires, it extinguishes by erasure.

Like any other machine, a Turing Machine is designed to *do* something. Confronting a set of symbols, its life is pretty much a matter of their transformation. It is thus a device of lunatic literalness, and indistinguishable, on this score, from the rest of humanity. At any given integral instant, a Turing Machine is capable of occupying one of a finite number of internal states: its *behavior* is completely determined when two temporal streams are fixed. In the first, the reading head changes its position on the tape; in the second, one state gives way to another. Having manipulated variously the symbols that it examines, a Turing Machine has transformed an *input* tape into an *output* tape.

There are certain concepts within mathematics that carry their significance on their wrist like a great tattoo. With other concepts, the significance takes some seeing. In what follows, I outline the construction of a specific Turing Machine capable of adding any two natural numbers. *Any* two, note. The machine's alphabet consists of the symbols 0 and 1. A natural number is represented by the machine as a string of $n + 1$ consecutive 1s—4 is thus simply 1 1 1 1 1. At any time, this machine may occupy one of six internal states, which I designate by means of the symbols $s_0$, $s_1$, $s_2$, $s_3$, $s_4$, and $s_5$. The symbol # indicates that a square on the tape is blank; L and R, that the machine is to move either to the left by one square or to the right. At HALT the machine stops. There are eight lines of instruction:

The first line has the following meaning: If the machine is in state $s_0$, and is examining a blank square, it is to print the symbol # on the square, and then move to state $s_0$. (It remains in the same state.) It then moves one square to the left. The remaining lines are read and understood in the same way.

Even the reader to whom mathematics is an affliction should appreciate the stunning power that mysteriously inheres in symbols. Here on the printed page, in a physical arrangement of ink (and that a concession only to the limitations of the *human* memory), is machinery enough to carry out the addition of any two numbers—stunning evidence, if any were needed, that the laws of thought owe little to the laws of physics, and represent, instead, a world in which concepts, like dreams, move entirely according to a logic of their own.

In the voluptuous generality of its conception, a Turing Machine is capable of standing in for any number of other concepts. From one point of view, it acts as a very model for the activity of computation; from another, it gives elegant content to Hilbert's notion of a formal system. But Turing's description of his Turing Machines also laid the foundations for the development of computer science, and so formed one of those landmarks in thought by which the past and the future are sharply divided.

Hilbert thought of mechanical reasoning in terms of a formal system whose elements were symbols and their rules. Turing thought in terms of an abstract machine. The formulations are quite different. The concepts are the same. Every formal system may be represented as a Turing Machine. Every Turing Machine embodies a formal system. In one of those inexplicable accidents of academic life, the American logicians Alonzo Church and Emil Post, following quite different lines of reasoning, defined a variety of abstract objects that in the end found expression as Turing Machines—an indication in such matters that the very same idea was being given inessentially different formulations. This prompted Church to argue that only one idea had been defined—*effective computability*.

It was Hilbert's idea to make mathematics mechanical by the double action of first withdrawing meaning from a set of symbols, and then reducing inference to a series of discrete, combinatorial steps. Gödel put an end to the Hilbert Program. Arithmetic is incomplete; its consistency cannot be demonstrated within arithmetic itself. This served to show that some truths must inevitably wriggle through any axiomatic net. Wriggling thus, they escape completely from mechanistic constraints. Gödel's theorem (and Tarski's theorem as well) established that the truth of an arithmetic proposition and its provability within a formal system are separate matters.

There yet remains a single, large-hearted question about the scope of purely mechanical concepts within mathematics. In the propositional calculus, the tautologies and the theorems coincide perfectly. This circumstance makes it possible to decide, with respect to an *arbitrary* proposition, whether it is a theorem. The mathematician need only establish whether it is a tautology. Having established *that*, a procedure exists for finding its proof. The problem in outline suggests the ping-ponging of a pair of questions: Is an arbitrary statement of a formal system a theorem? If it is, what (or where) is its proof? The questions taken together constitute the *decision problem* for a formal system.

In the third part to his (horribly unlucky) three-part program, Hilbert asked whether in general a decision procedure existed by which the decision problem could be settled. By a decision procedure, he meant an algorithm. That concept, of course, collapses into the concept of a formal system or a Turing Machine. In any event, to conclude on a negative note, the result of Turing's investigation was negative. Gödel showed that the truths of arithmetic did not all follow from the axioms of arithmetic. Turing demonstrated that the decision problem was insoluble as well. An element of thrilling and unassailable waywardness yet attaches to mathematical thought.

# 27. Life Itself

IN THE FIRST THIRTY YEARS OF THE TWENTIETH CENTURY, THE great physicists—Einstein, of course, Dirac, Schrödinger, Heisenberg, the taciturn Bohr—moved from triumph to triumph with the calm inevitability and special sense of grace afforded sleepwalkers in the dead of night. Of Einstein, especially, it was said that physics simply melted in his mouth. Others may see things differently, but to my mind the various branches of physics, whether quantum mechanics or relativistic astrophysics, appear to be converging toward a babbling form of mysticism in comparison with which the theory of the orgone box is a very marvel of clarity and sound good sense.

Einstein himself remarked late in life that the God of Physics—a figure, I imagine, much like Wotan—is subtle but not malicious. Einstein was vouchsafed an overwhelming vision of order early on; in the end, he acknowledged that what he had seen in a flash in the first flush of his young manhood was only a part of the truth, and he died in darkness, his work incomplete. This may not be evidence of malice in the scheme of things, but neither does it suggest much by way of divine helpfulness.

Thousands upon thousands of creatures swarm over the surface of the earth. Some are strikingly successful in the struggle for life. There are well over fifty thousand varieties of insects; like podiatrists, most of them are thriving. Some species have engaged in veritable prodigies of adaptation, with the hominids moving from an opposable thumb to a disposable tampon in what amounts to the blink of a biological eye. Others, like the shark, never adapt to anything new at all and represent only a persistent cough in the long dark night of evolution. Living creatures grow, and they carry on ceaselessly, striving, acquiring, laying up provision and storing it, cunning and resourceful; they reach out, impelled by some dim ancestral power, some muted voice heard, with a musky throb of recognition by everything that lives, one to another; they mate and multiply, and then, unreconciled and, for all I know, unreconcilable, they sink into death, decay, and oblivion.

More than any particular thing that lives, life itself suggests a kind of intelligence evident nowhere else; reflective biologists have always known that in the end they would have to account for its fantastic and controlled complexity, its brilliant inventiveness and diversity, its sheer *difference* from anything else in this or any other world.

# ENDNOTES

## INTRODUCTION

1. David Chalmers, "Is the Soul Immortal?," interview by Robert Kuhn, *Closer to the Truth*, YouTube, May 4, 2021, video, 9:06, https://www.youtube.com/watch?v=Bejm1mYsr5s&t=34s.
2. Lisa Randall, *Warped Passages: Unraveling the Mysteries of the Universe's Hidden Dimensions* (New York: Harper Perennial, 2006), 158.
3. Blaise Pascal, *Pensées* (1670).
4. Samuel Johnson to Lord Chesterfield (Philip Dormer Stanhope), February 7, 1755, https://jacklynch.net/Texts/letter55.html.

## 2. MISPRINTS IN THE BOOK OF LIFE

1. There are additional nuances. For instance, a twenty-first amino acid, selenocysteine, containing selenium, appears less frequently and codes and translates differently from the canonical twenty amino acids. For more on the outlier, see Robert Longtin, "A Forgotten Debate: Is Selenocysteine the 21st Amino Acid?," *Journal of the National Cancer Institute* 97, no. 7 (April 7, 2004): 504–505, https://doi.org/10.1093/jnci/96.7.504.

## 4. THE EVIDENCE FOR EVOLUTION

1. As for the peppered moth experiment, it turns out to have been riddled with problems. And as biologist Jonathan Wells notes, "Even if the classic peppered moth story were true, it would not confirm Darwin's claim that the new species, organ, and body plans were produced by unguided evolution. All it would demonstrate is that natural selection produced a shift in the proportions of two existing varieties of the same species." Jonathan Wells, *Zombie Science: More Icons of Evolution* (Seattle, WA: Discovery Institute Press, 2017), 64. In the fruit fly experiment, artificial irradiation—introduced to accelerate the rate of genetic mutations—did lead to morphological changes, but in every case the changes were either damaging or lethal to the ill-fated offspring. For more on the problems with the fruit fly and peppered moth experiments, see Chapter 3 of *Zombie Science* and Jonathan Wells, *Icons of Evolution: Why Much of What We Teach about Biology is Wrong* (Washington, DC: Regnery, 2000), Chapters 7 and 9.
2. David B. Kitts, "Paleontology and Evolutionary Theory," *Evolution* 28, no. 3 (September 1974): 467.

3. Ernst Mayr, *Population, Species, and Evolution* (Cambridge, MA: Belknap Press, 1970), 107, 117.

4. Albert Jacquard, *Biomathematics: The Genetic Structure of Populations, Vol 5* (New York: Springer-Verlag, 1974), 271.

## 6. DARWIN AND THE MATHEMATICIANS

1. John von Neumann quoted in Lily E. Kay, *Who Wrote the Book of Life?: A History of the Genetic Code* (Stanford, CA: Stanford University Press, 2000), 158. Kay cites von Neumann to Gamow, July 25, 1955, box 4, Gamow folder, von Neumann Papers, LC.

2. Von Neumann quoted in *Who Wrote the Book of Life?*, 158.

3. Douglas L. Theobald, *Panda's Thumb*, August 20, 2008, https://pandasthumb.org/archives/2008/08/von-neumann-on.html.

## 8. A LONG LOOK BACK: THE WISTAR SYMPOSIUM

1. The proceedings of the conference were later published as *Mathematical Challenges to the Neo-Darwinian Interpretation of Evolution*, eds. Paul S. Moorhead and Martin M. Kaplan (Philadelphia: The Wistar Institute Press, 1967).

2. I talked several times with Murray Eden about the views that he expressed at the Wistar Symposium, but never with the kind of intellectual intimacy that I enjoyed with M. P. Schützenberger. My recollection of our many discussions, and these over the course of ten years, or more, are based in large parts on the notebooks I kept of these discussions.

3. This word is, of course, a *sentence* in English.

4. H. A. Prichard, "Does Moral Philosophy Rest on a Mistake," [1912] *Moral Writings*, ed. Jim MacAdam (Oxford, UK: Clarendon Press, 2002), 12.

5. René Thom, *Esquisse d'une sémiophysique: physique aristotélicienne et théorie des catastrophes* (Paris: InterÉditions, 1988), 53. Emphasis in original.

6. In this respect, see Marc Henry, "The Hydrogen Bond," *Inference* 1, no. 2 (March 2015), https://inference-review.com/article/the-hydrogen-bond.

7. Thom had the idea; John Mather and Bernard Malgrange provided the proof. For an outline of the proof, see my own *Black Mischief: Language, Life, Logic & Luck* (New York: Harvest Books, 1988).

8. An abbreviated version of Michael Denton's book may be found in three installments in the first three issues of *Inference*: part one, *Inference* 1, no. 1 (October 2014), https://inference-review.com/article/evolution-a-theory-in-crisis-revisited-part-one; part two, *Inference* 1, no. 2 (March 2015), https://inference-review.com/article/evolution-a-theory-in-crisis-revisited-part-two; and part three, *Inference* 1, no. 3, https://inference-review.com/article/evolution-a-theory-in-crisis-revisited-part-three.

9. D. L. Hartl, "Evolving Theories of Enzyme Evolution," *Genetics* 122, no. 1 (May 1989): 1, https://doi.org/10.1093/genetics/122.1.1.

10. The example is taken from Tyler Hampton's review of Dan Tawfik's laboratory in *Inference* 1, no. 1 (October 2014), https://inference-review.com/article/the-new-view-of-proteins.

11. Mónica R. Buono, Marta S. Fernández, Marcelo A. Reguero, Sergio A. Marenssi, Sergio N. Santillana, and Thomas Mörs, "Eocene Basilosaurid Whales from the Las Meseta

Formation, Marambio (Seymor) Island, Antarctica,"*Ameghiniana* 53, no. 3 (2016): 311, https://doi.org/10.5710/AMGH.02.02.2016.2922. The authors attempt to restore an 8–10 million year window for the evolutionary process via an alternate reading of the geological evidence, but as Jonathan Wells argues, the move smacks of special pleading and, in any case, even 8–10 million years appears far from sufficient once the problem of combinatorial inflation is squarely faced. Jonathan Wells, *Zombie Science: More Icons of Evolution* (Seattle, WA: Discovery Institute Press, 2017), 111–114.

12. A. Carbone and M. Gromov, "Mathematical Slices of Molecular Biology," January 17, 2001, Institut des Hautes Études Scientifiques, IHMES/M/01/03, https://www.ihes .fr/~gromov/wp-content/uploads/2018/08/M01-03.pdf. Emphasis in original. Internal references removed.

13. Caleb Scharf, "Is Physical Law an Alien Intelligence?," *Nautilus*, November 11, 2016, https://nautil.us/is-physical-law-an-alien-intelligence-236218/.

## 10. The Activity of a Cell Is Like That of a Factory

1. Edward R. Dougherty, "Science without Validation in a World without Meaning," *American Affairs* 4, no. 2 (Summer 2020): 90–106, https://americanaffairsjournal.org /2020/05/science-without-validation-in-a-world-without-meaning/. Emphasis added. Internal references removed.

## 11. A Graduate Student Writes

1. Nick Matzke, "Meyer's Hopeless Monster, Part II," *Panda's Thumb*, June 19, 2013, http://pandasthumb.org/archives/2013/06/meyers-hopeless-2.html.

2. J. P. Lin et al., "A *Parvancorina*-like Arthropod from the Cambrian of South China," *Historical Biology* 18, no. 1 (2006): 33–45.

3. I took this quoted passage from the Trilobite entry at Wikipedia (in July 2013), a Party Organ and hardly a source our side is disposed to champion. At some point in the in-tervening decade, the Wikipedia page was modified to remove the imprudently candid admission, but as of this writing the passage lives on at another site that borrowed heavily from the pre-airbrushed Wikipedia entry (and gives due credit for such): the entry for Trilobite at *Geology Page*, November 14, 2013, https://www.geologypage.com/2013/11 /trilobite.html ( accessed January 17, 2023). Why stop there? See also P. Jell, "Phylogeny of Early Cambrian Trilobites," in R. A. Fortey, P. D. Lane, and D. J. Siveter, *Trilobites and Their Relatives: Contributions from the Third International Conference*, Oxford 2001, Special Papers in Palaeontology, no. 70 (London: Palaeontological Association, 2003): 45–57.

4. "The Cambrian Period," UC Museum of Paleontology, last modified July 6, 2011, https://ucmp.berkeley.edu/cambrian/cambrian.php.

## 12. A One-Man Clade

1. Günter P. Wagner, *The Character Concept in Evolutionary Biology* (San Diego, CA: Aca-demic Press, 2000), xv.

2. Nick Matzke, "Meyer's Hopeless Monster, Part II," *Panda's Thumb*, June 19, 2013, http://pandasthumb.org/archives/2013/06/meyers-hopeless-2.html.

3.  Keynyn Brysse, "From Weird Wonders to Stem Lineages: The Second Reclassification of the Burgess Shale Fauna," *Studies in History and Philosophy of Biological and Biomedical Sciences* 39 (2008): 298–313.

4.  David Legg et al., "Cambrian Bivalved Arthropod Reveals Origin of Arthrodization," *Proceedings of the Royal Society, Series B: Biological Sciences* 279, no. 1748 (2012): 4699–4704.

5.  Willi Hennig, *Grundzüge einer Theorie der phylogenetischen Systematik* (Berlin: Deutscher Zentralverlag, 1950). No real cladist would ever refer to the English translation of this book. Let me see: Matzke never once refers to the German original. Readers must draw their own conclusions.

6.  A. V. Z. Brower, "Evolution Is Not a Necessary Assumption of Cladistics," *Cladistics* 16 (2000): 143–154. Later in his paper, Brower remarks that "systematics provides evidence that allows inference of a scientific theory of evolution." What doesn't?

7.  See G. P. Wagner and P. F. Stadler, "Quasi-independence, Homology and the Unity of Type: A Topological Theory of Characters," *Journal of Theoretical Biology* 220 (2003): 505–527.

8.  Matzke, "Meyer's Hopeless Monster."

9.  M. Schmitt, "Claims and Limits of Phylogenetic Systematics," *Journal of Zoological Systematics and Evolutionary Research* 27 (1989): 181–190.

10. A point made vividly by Matzke's own source, which he cites in solemn incomprehension: Whatever the character matrix, Keynyn Brysse observes, "there is only enough reliable information available to construct cladograms, not trees." Brysse, "From Weird Wonders to Stem Lineages," 306.

11. Schmitt, "Claims and Limits of Phylogenetic Systematics," Schmitt was for many years the curator of Coleoptera and Head of Department of Arthropoda at the Zoologisches Forschungsmuseum Alexander Koenig. What he does not know about beetles is apparently not worth knowing.

12. "It is easy to calculate statistics such as CI and RI," Matzke writes, "and compare them to CI and RI statistics calculated based on data reshuffled under a null hypothesis where any possible phylogenetic signal has been obliterated." True enough. It is easy. "In virtually any real case, one will see substantial phylogenetic signal, even if there is uncertainty in certain portions of the tree." True enough again. The question of what the signal is signaling remains.

13. Matzke, "Meyer's Hopeless Monster."

14. Stephen Jay Gould, *The Structure of Evolutionary Theory* (Cambridge, MA: The Belknap Press of Harvard University Press, 2002), 1057.

15. Matzke, "Meyer's Hopeless Monster."

16. Gregory D. Edgecombe, "Arthropod Phylogeny: An Overview from the Perspective of Morphology, Molecular Data, and the Fossil Record," *Arthropod Structure & Development* 39 (2010): 74–87. Edgecombe is not alone. The characters Matzke thinks homologous, Legg calls into question. See the caption for Figure 3, Legg, "Cambrian Bivalved Arthropod." Legg is Matzke's witness and not ours. Valentine and Erwin do as much. Another witness, this one expert: "The lobopodians all share fairly simple, unspecialized legs, yet *Opabinia* and anomalocaridids lack legs but have paired, lateral flaps that, particularly in *Opabinia*, have gills along the upper aspect of the flap. Beyond the Radionta, however, well-sclerotized jointed appendages reappear. Are arthropod appendages

homologous to those of lobopods, as Budd has argued? Are they homologous to the lateral flaps of Radionta [the group that includes anomalocaridids]? Or are they entirely novel structures? This debate is far from settled, illustrating the complexities of understanding the evolutionary pathways among these groups." Douglas Erwin and James Valentine, *The Cambrian Explosion: The Construction of Animal Biodiversity* (Greenwood Village, CO: Roberts and Company, 2013), 195. (Internal citations removed.) Vengeance is mine, saith the Lord.

17. Matzke, "Meyer's Hopeless Monster." Emphasis in original.

## 13. Good as Gould

1. Stephen Jay Gould, *Ever Since Darwin: Reflections on Natural History* (New York: W. W. Norton & Company, 1977), 26.

2. Gould, *Ever Since Darwin*, 27.

## 15. A Natural History of Curiosity

1. For context and a slightly different translation of the passage, see Bediuzzaman Said Nursi, *The Flashes Collection*, trans. Şükran Vahide (Istanbul: Sözler Neşriyat A.Ş., 2009), 222, https://archive.org/stream/RisalaNur/TheFlashes_djvu.txt.

2. "A Letter of Jamshid al-Kâshi to His Father," trans. E. S. Kennedy, in *Studies in the Islamic Exact Sciences*, eds. David A. King and Mary Helen Kennedy (Beirut, Lebanon: American University of Beirut, 1983), 724–744.

3. Louis-Amélie Sédillot, *Prolégoménes des Tables Astronomiques d'Oloug-Beg: Traduction et Commentaire* (Paris: Firmin Didot Fréres, 1853), 4. Translated from the French by David Berlinski.

4. Aydin Sayili, *The Observatory in Islam and Its Place in the General History of the Observatory* [1960], 2nd ed. (Ankara, Turkey: Turk Tarih Kurumu Basimevi, 1988), 391.

5. Sayili, *The Observatory in Islam*, 11.

6. Sayili, *The Observatory in Islam*, 11.

7. Sayili, *The Observatory in Islam*, 11.

8. Bernard of Clairvaux, "Letter 193," in *For and Against Abelard: The Invective of Bernard of Clairvaux and Berenger of Poitiers*, eds. R. M. Thomson and M. Winterbottom (Rochester, NY: Boydell & Brewer, 2020), 35.

9. Al-Ghazali, *The Incoherence of the Philosophers* (*Tahāfut al-Falāsifah*): *A Parallel English-Arabic Text*, trans. Michael E. Marmura (Provo, Utah: Brigham Young University Press, 2000), 166. Bracketed insertions in translation.

10. David Hume, "Enquiry Concerning Human Understanding," in Terence Penelhum, *David Hume: An Introduction to His Philosophical System* (West Lafayette, IN: Purdue University Press, 1992): 76.

## 16. The Ineffable Higgs

1. Sean Carroll, "After the Higgs Boson: What Scientists Will Do with the Discovery," *Daily Beast*, July 6, 2012, updated July 13, 2017, https://www.thedailybeast.com/after-the-higgs-boson-what-scientists-will-do-with-the-discovery.

2. Lawrence M. Krauss, "How the Higgs Boson Posits a New Story of Our Creation," *Newsweek*, July 9, 2012, https://www.newsweek.com/how-higgs-boson-posits-new -story-our-creation-65567.

3. Krauss, "How the Higgs Boson Posits a New Story of Our Creation."

4. Ptolemy, *Almagest*, trans. G. J. Toomer (Princeton, NJ: Princeton University Press, 1998), 38.

5. Sean Carroll, "Hidden Symmetries," *Discover Magazine*, October 24, 2005.

6. Art Hobson, "There Are No Particles, There Are Only Fields," submitted for publication to *American Journal of Physics*, April 18, 2012, 17, https://dokumen.tips/documents /there-are-no-particles-there-are-only-fieldspdf.html?page=17.

7. Hobson, "There Are No Particles, There Are Only Fields," 1.

8. Steven Weinberg, "What Is Quantum Field Theory, and What Did We Think It Is?," *Arxiv*, February 4, 1997, https://doi.org/10.48550/arXiv.hep-th/9702027 (lecture, Historical and Philosophical Reflections on the Foundations of Quantum Field Theory conference, Boston University, March 1996).

9. Hobson, "There are No Particles," 1.

## 17. THE GOOD SOLDIER

1. Ecclesiastes 1:1–2.

2. Brian Greene, *Until the End of Time: Mind, Matter, and Our Search for Meaning in an Evolving Universe* (UK: Penguin Books, 2020), 16.

3. Greene, *Until the End of Time*, 306–309. One of the advantages of the multiverse is just that it provides an answer of sorts to various fine-tuning questions. Why are the laws of physics what they are or the fundamental constants as they are? In some other universe, they are otherwise. As luck would have it, we live in a universe where the laws of physics are what they are and the fundamental constants as they are. So far Greene is bound to go in spirit, but on page 307, the conservation of energy is promoted to duty *across* the multiverse. This might suggest that the fundamental laws of physics are necessary, and so true in all possible worlds. This answers a question that Greene raises on page 52: "Why *these* laws instead of *those*?" [Emphasis in original.] Were it not for the fact that the laws of physics look nothing like necessary propositions, this would be a very elegant view. It is also a view that would strengthen Greene's argument against free will. The laws of physics do nothing to prevent a man from acting freely. That is not their business. But they reveal that free action is impossible in the same sense that there is no violating the principle that everything is identical to itself. But this supposes that there is a *conflict* between the laws of nature and the existence of free will. No one has succeeded in demonstrating the conflict. If I lift my hand, what law of nature do I violate? And yet there is that stubborn difference, as Wittgenstein observed, between lifting my hand and the fact that my hand goes up. This is a part of the problem of free will.

4. Greene, *Until the End of Time*, 118.

5. If Newtonian mechanics is time-reversible, Joseph Löfschmidt observed, there is no obvious way in which to derive the second law of thermodynamics from its assumptions. Ludwig Boltzmann's H-theorem demonstrated that in progressing from a non-equilibrium to an equilibrium state, a physical system must increase in entropy. The argument is fine. Its premises hinge on the assumption of molecular chaos, or

*Stosszahlansatz*, which cannot be derived from Newtonian mechanics itself. If this is so, then neither can the second law of thermodynamics. In this regard, see Jochen Gemmer, Alexander Otte, and Gunter Mahler, "Quantum Approach to a Derivation of the Second Law of Thermodynamics," arXiv, January 20, 2001, https://arxiv.org/abs/quant-ph/0101140, for references, as well as Barbara Drossel, "On the Relation between the Second Law of Thermodynamics and Classical and Quantum Mechanics," arXiv, August 27, 2014, https://arxiv.org/abs/1408.6358. Witness the fearless Drossel: "In the following, I will argue that the second law of thermodynamics cannot be derived from deterministic time-reversible theories such as classical or quantum mechanics." In this, Drossel is on her own.

6. Greene, *Until the End of Time*, 118.

7. Greene, *Until the End of Time*, 118.

8. Greene, *Until the End of Time*, 9.

9. Writing in the September 9, 2009, issue of *Scientific American*, Ernst Mayr remarked that "biological concepts, and the theories based on them, cannot be reduced to the laws and theories of the physical sciences." The point is obvious. No theory of particle physics makes reference to random variation or natural selection; nor do theories of evolution appeal to quark confinement or the Higgs boson. Alex Rosenberg, a philosopher of science, has been led by this consideration to include the principle of natural selection among the fundamental laws of physics. It is a view that John Dupre modestly described as implausible. See John Dupre, "Is Biology Reducible to the Laws of Physics," *The American Scientist*, May-June, 2007, https://doi.org/10.1511/2007.65.276. Chiara Marletto discusses the same questions from the point of view of what she calls constructor theory. See, in particular, her essay, "Life without Design," in *Aeon*, July 2015, https://aeon.co/essays/how-constructor-theory-solves-the-riddle-of-life. My own point of view, acquired many years ago from Pat Suppes, my colleague at Stanford in the good old days, is that model theory is the proper discipline in which to express the philosophy of science. See Newton Da Costa and Steven French, "The Model-Theoretic Approach in the Philosophy of Science," *Philosophy of Science* 57, no. 2 (June 1990): 248–265. There is something that this approach—my approach—never addresses, and that is the mystical nature of the fundamental laws of physics.

10. Greene, *Until the End of Time*, 237. Emphasis in original.

11. Michael Peskin and Daniel Schroeder, *An Introduction to Quantum Field Theory* (Boulder, CO: Westview Press, 1995).

12. See Francis J. Kovach, "The Enduring Question of Action at a Distance in Saint Albert the Great," *The Southwestern Journal of Philosophy* 10, no. 3 (1979): 161–235, www.jstor.org/stable/43155503.

13. Greene, *Until the End of Time*, 146.

14. The point carries over to mathematics. The proposition that two plus two equals four does not, of course, *make* two plus two equal four. Numbers, like elementary particles, do not answer to our theorems.

15. Greene, *Until the End of Time*, 146–159.

16. Greene, *Until the End of Time*, 147. There is nothing modern about this argument.

17. Let alone knocked up.

18. William Shakespeare, *King Lear*, 4.1.41–42.

19. Ecclesiastes 9:11, KJV.

20. Greene, *Until the End of Time*, 149.

21. It is not obvious that a man is free to the extent that he could have done otherwise. *Here I stand*, Martin Luther declared, *I can do no other*. What judgment about freedom of the will follows?

22. Greene, *Until the End of Time*, 149.

23. Greene, *Until the End of Time*, 132 and following. In these pages, Greene describes a number of popular theories about consciousness and information, all of them richly preposterous.

24. These issues owe much to Thomas Nagel's influential paper, "What Is It Like to Be a Bat?," *The Philosophical Review* 84, no. 4 (October 1974): 435–450. What *is* it like to be a bat? One answer is obvious. To be a bat is like being me were I given to squeaking loudly and hanging by my feet. This description might be suitably enlarged to encompass eating insects and navigating by sonar. On the other hand, we might ask what it is like for a *bat* to be a bat? And the obvious answer is, who knows? But the obvious answer has nothing to do with bats. I can no more imagine what it is like for a bat to be a bat than I can imagine what it is like for anyone but me to be anyone but me. That is a part of the nature of consciousness: to each his own. Nagel concludes that some experiences cannot be described by any proposition; and from this it follows that a physical description of the world must be incomplete. This seems to me a mistake. If what is derived from a theory is a prediction, it cannot be an experience, and if it is an experience, it cannot be derived from a theory. The same mistake is at work in Frank Jackson's essay, "What Mary Didn't Know," *Journal of Philosophy* 83, no. 5 (May 1986): 291–295, https://doi.org/10.2307/2026143. What we are left with is the uninspiring conclusion that I cannot have your experiences and vice versa. No doubt, we are both well pleased.

25. W. Somerset Maughan, *Of Human Bondage* (New York: George H. Doran Company, 1915), Chapter 57.

26. See Toby Cubitt, David Perez-Garcia, and Michael M. Wolf, "Undecidability of the Spectral Gap," arXiv, April 18, 2018, https://arxiv.org/abs/1502.04573v3. There are a surprising number of undecidable propositions within mathematical physics, evidence, if any were needed, that undecidability is not simply an oddity of elementary number theory.

27. Ptolemy, "On the Qualities of the Soul," *Tetrabiblos*, III.8. See F. E. Robbins, *The Tetrabiblos* (Cambridge, MA: Harvard University Press, 1980).

28. Greene writes often that if A contains B, or is otherwise made up of B-like elements, then the laws that explain B must explain A as well. This is a little like saying that if a bag contains chocolate chip cookies, their recipe must explain something about the bag. In the end, this kind of appeal to the constituents of things represents little more than an attempt to reify the relationship between theories. The requisite diagram does not commute. There is no such relationship in nature.

29. William Shakespeare, *The Tempest*, 5.1.59–66.

## 20. A Review of Michael Ruse's *The Philosophy of Biology*

1. Michael Ruse, *The Philosophy of Biology* (London: Hutchinson University Library, 1973), 33.

2. Ruse, *The Philosophy of Biology*, 37.

3.  E. C. Zeeman, "Differential Equations for the Heartbeat and the Nerve Impulse" in *Towards a Theoretical Biology, Vol.4*, ed. C. H. Waddington (Chicago: Aldine-Atherton, 1972).

4.  Ruse, *The Philosophy of Biology*, 69.

5.  Ruse, *The Philosophy of Biology*, 97.

6.  Ernst Mayr, *Population, Species, and Evolution* (Cambridge: Belknap Press, 1970), 107.

7.  Mayr, *Population, Species, and Evolution*, 117.

8.  William Feller, "On Fitness and the Cost of Natural Selection," *Genetical Research* 9 (1967): 1–15, https://www.cambridge.org/core/services/aop-cambridge-core/content/view/B43FA212FD8A0812F7F324F90C80B018/S0016672300010260a.pdf/on_fitness_and_the_cost_of_natural_selection.pdf.

9.  A. R. Manser, "The Concept of Evolution," *Philosophy* 40, no. 151 (1965): 18–34.

10. See, for example, *Mathematical Challenges to the Neo-Darwinism Interpretation of Evolution*, eds. Paul S. Moorhead and Martin M. Kaplan (Philadelphia: Wistar Institute Press, 1967); and Berlinski, "Philosophical Aspects of Molecular Biology," *The Journal of Philosophy* (June 1972).

11. Bernhard Rensch, "The Laws of Evolution," in *Evolution after Darwin*, ed. Sol Tax (Chicago: University of Chicago Press, 1960).

12. Marcel Schützenberger, "Algorithms and the neo-Darwinian Theory of Evolution," in *Mathematical Challenges to the Neo-Darwinism Interpretation of Evolution*, eds. Paul S. Moorhead and Martin M. Kaplan (Philadelphia: Wistar Institute Press, 1967).

13. Ruse, *The Philosophy of Biology*, 153.

14. Ruse, *The Philosophy of Biology*, 153.

15. Ruse, *The Philosophy of Biology*, 153.

16. Ruse, *The Philosophy of Biology*, 153.

17. H. H. Ross, "Review of Principles of Numerical Taxonomy," *Systematic Zoology* 13 (1964): 106–108.

## 21. THE DIRECTOR'S CUT

1.  Original: "*Creatio numerorum, rerum est creatio.*" Thierry of Chartres, *The Commentary on the* De arithmetica *of Boethius*, ed. Irene Caiazzo (Toronto: Pontifical Institute of Mediaeval Studies, 2015), ix.

2.  Colin McLarty has even suggested that it may be possible to derive Fermat's last theorem from within Peano arithmetic. If true, this would be remarkable. See Colin McLarty, "The Large Structures of Grothendieck Founded on Finite Order Arthimetic," arXiv, February 2011, revised April 2014, https://arxiv.org/abs/1102.1773v4.

3.  Kurt Gödel, "Über formal unentscheidbare Sätze der Principia Mathematica und verwandter Systeme I," *Monatshefte für Mathematik und Physik* 38 (1931): 173–198. The English translation of this masterpiece appeared almost thirty years after its publication in German: *On Formally Undecidable Statements of* Principia Mathematica *and Related Systems*, trans. Bernard Meltzer (New York: Basic Books, 1962).

4.  Ernst Zermelo thought it contained a mistake and could not be persuaded otherwise. Paul Finsler insisted that he had anticipated Gödel's result, vexing Gödel endlessly with claims of priority. Emil Post, a very great logician, *had* anticipated Gödel's result and by

ten years—Gödel's result, but not his methods. See John Dawson, Jr., "The Reception of Gödel's Incompleteness Theorems," *PSA: Proceedings of the Biennial Meeting of the Philosophy of Science Association, Vol. II: Symposia and Invited Papers* (1984): 253–271.

5.  Logicians now say that they take the theorem and its proof in stride. No doubt. They have had ninety years to study it. But, while Gödel's proof is no longer mystifying, it remains what it always was and that is rebarbative.

6.  Gödel, *On Formally Undecidable Statements*, 38. The translation in the more recent edition published by Solomon Feferman et al. has "in outward appearance," which is both unidiomatic (as opposed to "to all outward appearances"), and pointlessly passive. The German quite plainly means "considered (*betrachet*) from the outside (äusserlich)." Someone is considering something. Feferman et al. remark that Gödel approved their translation. This is no very great recommendation. Gödel was not a native English speaker. Stefan Bauer-Mengelberg, who worked closely with Gödel as his translator, once remarked to me that Gödel was infuriating in his obsessive commitment to his own idiosyncratic version of English grammar. See Kurt Gödel, *Collected Works, Vol. I*, eds. Solomon Feferman et al. (New York: Oxford University Press, 1989), 147.

7.  German, of course, in the original.

8.  Clement Greenberg remarks, "The Old Masters had sensed that it was necessary to preserve what is called the integrity of the picture plane: that is, to signify the enduring presence of flatness underneath and above the most vivid illusion of three-dimensional space. The apparent contradiction involved was essential to the success of their art, as it is indeed to the success of all pictorial art. The Modernists have neither avoided nor resolved this contradiction; rather, they have reversed its terms. *One is made aware of the flatness of their pictures before, instead of after, being made aware of what the flatness contains* [emphasis added]. Whereas one tends to see what is in an Old Master before one sees the picture itself, one sees a Modernist picture as a picture first. This is, of course, the best way of seeing any kind of picture, Old Master or Modernist, but Modernism imposes it as the only and necessary way, and Modernism's success in doing so is a success of self-criticism." Clement Greenberg, "Modernist Painting" [1965], reprinted in *Modern Art and Modernism*, eds. Francis Francina and Charles Harrison (New York: Harper & Row, 1982), http://www.columbia.edu/itc/barnard/arthist/wolff/pdfs/week4_greenberg.pdf.

9.  More of the same in the sense that 'F' in 'F($x$)' is itself designated a variable, although not one open to quantification. Were it open, the system that results would be second order.

10. Latvian, as it happens.

11. It is crystal clear as an inference from P ⊃ P and P to P. Trivial? Of course. But crystal clear anyway.

12. And the logical axioms of Alfred North Whitehead and Bertrand Russell's *Principia Mathematica*.

13. Feferman remarks of formal systems that their "formulas are built up from basic arithmetical (and possibly other) relations by means of the propositional connectives (such as ¬, ∧, ∨ →) and quantifiers (such as ∀, ∃) and whose provable formulas are obtained from a given set of axioms (both logical and non-logical) by closing under certain rules of inference. Moreover, F is assumed to be effectively given, i.e. the set of axioms of F and its rules of inference are supposed to be effectively decidable, so that its set of provable formulas is effectively enumerable." Nice. Solomon Feferman, "Penroses's Gödelian Argument," *Psyche* 2, no. 7 (May 1995), https://citeseerx.ist.psu.edu/document?repid=rep1&type=pdf&doi=52f06b3d2f68a79378311e44a5b399f06d9c6ef6.

14. Samuel Johnson, *A Journey to the Western Islands of Scotland* [1775], quoted in *Oxford Essential Quotations*, ed. Susan Ratcliffe (Oxford: Oxford University Press, 2017).

15. Gödel, *On Formally Undecidable Statements*, 38.

16. Whenever it is clear in context whether a formula is being named or used, quotation marks are dropped.

17. This is Newspeak; Gödel's original term was *rekursiv*. The distinction between primitive recursion and recursion is important. Hence Newspeak.

18. Gödel, *On Formally Undecidable Statements*, 46.

19. Division, no, but addition and multiplication, yes. It is possible to define addition and multiplication recursively within the Peano axioms, but the definitions require the recursion theorem, and this, in turn, requires a certain amount of set theory.

20. Gödel, *On Formally Undecidable Statements*, 49.

21. Proofs are recursively enumerable, but not recursive; truths are neither.

22. Gödel did not prove the fifth theorem, leaving his remarks as a sketch. See Gödel, *On Formally Undecidable Statements*, 55–56.

23. Simple consistency suffices for the first claim; ω-consistency for the second. Shortly after Gödel published his work, J. Barkley Rosser showed how to demote ω-consistency in favor of simple consistency all over again.

24. The variable *sign* '*x*' giving way to the numeral.

25. Gödel, *On Formally Undecidable Statements*, n. 10.

26. The essential argument is now called the fixed-point lemma, or the diagonalization lemma, and may be expressed with less by way of elaboration. Let $A(x)$ designate a formula in $\text{FP}_A$ with one free variable in $x$. It is always possible to construct a formula **G**, designating some formula in $\text{FP}_A$ such that $\text{FP}_A \vdash \mathbf{G} \ll A(n)$, where $n$ is the Gödel number of **G**, and $n$ the numeral naming it in $\text{FP}_A$. On the level of the underlying assembly language, $n$ is written as $S, \ldots, S(0)$. With the identification of A in $A(x)$ as ~**Bew**, it follows that $\text{FP}_A \vdash \mathbf{G} \ll \mathbf{G}(n)$, with **G** saying of itself that it is unprovable. This is, perhaps, a simpler way of presenting things. Gödel's introduction is rhetorically more effective than contemporary accounts because it suggests, if only indirectly, some of the heavy lifting that goes into the full formal proof. The claim that $[\alpha; n]$ is the formula that results when $x$ is replaced by $n$ in $\alpha$ at once invites two questions: Just what results? And how is the replacement carried out? The requisite operation is one of substitution, and it is no easy business to define it properly. No matter the rhetorical ease afforded by Gödel's introduction, some logicians considered it a mistake because it replaces consistency with soundness. The development of model theory has made this argument anachronistic.

27. It is here that Gödel's introduction, in appealing to soundness rather than consistency, parts company with his proof.

28. Alfred Tarski, "The Concept of Truth in Formalized Languages (1935)," in *Logic, Semantics, Metamathematics*, ed. John Corcoran, trans. Joseph Woodger (Indianapolis: Hackett, 1983), 152–278, whence the concept of truth. See Alfred Tarski and Robert Vaught, "Arithmetical Extensions of Relational Systems," *Compositio Mathematica* 13 (1956): 81–102, for the definition of the far more elegant concept of truth in a model.

29. A relation R of $n$ arguments on D is said to be **M**-definable *with parameters* if it is the extension in **M** of some formula of $\text{FP}_A$ containing $m + n$ free variables, $m$ of which have been fixed in D.

30. S is a sentence and, by itself, it cannot code for anything. Nonetheless, it is easy to show that it is equivalent to some formula A(x) with one free variable whose translate within the system takes over the coding.

31. John Lucas, "Mind, Machines and Gödel," *Philosophy* 36 (April–July 1961): 112–27.

32. Lucas, "Mind, Machines and Gödel," 113.

33. Roger Penrose, *Shadows of the Mind* (New York: Oxford University Press, 1994), 65.

34. Hilary Putnam, "Minds and Machines," in *Dimensions of Mind*, ed. Sidney Hook (New York: New York University Press, 1960), 138–64.

35. Lucas, in rebuttal, has not covered himself in glory. Addressing an imaginary opponent, he remarks that "if, however, he acknowledges that the system cannot prove its Gödelian formula, then we know it is consistent, since it cannot prove every well-formed formula, and knowing that it is consistent, know also that its Gödelian formula is true." But this is to collapse the discussion into the frank fallacy in which P ⊃ Q and Q give rise to P.

36. It is perfectly possible to see that still waters run deep; *and* to see nothing on seeing *mierīgie ūdeņi ir tie dziļākie; and* yet to know perfectly well that they are in some sense the same.

37. Panu Raatikainen in a review of Torkel Franzen's "Gödel's Theorem: An Incomplete Guide to Its Use and Abuse," in *Notices of the American Mathematical Society* 54, no. 3 (March 2007): 382.

38. Kurt Gödel, *Collected Works, Vol. I: Publications 1929–1936*, eds. Solomon Feferman et al. (New York: Oxford University Press, 1989), 151. The German is *richtig*.

39. FP$_A$ says nothing about truth either as a relation or a formal sign, and, in any case, Tarski's theorem makes it plain that this claim is untrue as stated. Other systems may admit a partial definition of **Tr**, or a set of axioms incorporating **Tr**, but *they*, of course, are not *this* system, our very own.

40. Kurt Gödel, "Some Basic Theorems on the Foundations of Mathematics and Their Implications (1951)," in *Collected Works, Vol. III: Unpublished Essays and Lectures*, eds. Solomon Feferman et al. (New York: Oxford University Press, 1995), 310.

41. Gödel quoted in Hao Wang, *A Logical Journey: From Gödel to Philosophy* (Cambridge, MA: MIT Press), 207.

42. For the moment, we possess no real theories about human nature or human life: we are immersed in the flow. Whatever the theories we may in the future have, there is no obvious reason to suppose that these theories and our theories of the physical world should be consistent, and every reason to suppose the contrary. This means only that they cannot be simultaneously upheld, reason enough not to uphold them simultaneously.

## 22. Comments on Stuart Pivar's *Lifecode*

1. Stuart Pivar, *Lifecode: The Theory of Biological Self-Organization* (New York: Ryland Press, 2004).

2. René Thom, *Structural Stability and Morphogenesis*, trans. D. H. Fowler (Reading, MA: W. A. Benjamin, 1975), 8.

## 23. Catastrophe Theory and Its Applications: A Critical Review

1. René Thom, *Structural Stability and Morphogenesis* [1972] (Reading, MA: Benjamin, 1975).

2. Christopher E. Zeeman, *Catastrophe Theory* (Reading, MA: Addison-Wesley, 1977). Thom responds to Zeeman in *Dynamical Systems—Warwick 1974: Proceedings of a Symposium Held at the University of Warwick 1973/74*, ed. Anthony Manning (New York: Springer-Verlag, 1975), 384–389.

3. Hector J. Sussmann and Raphael S. Zahler, "Catastrophe Theory as Applied to the Social and Biological Sciences: A Critique," *Synthese* 37, no. 2 (1978): 117–217.

4. J. Mather, "Stratifications and Mappings," in *Dynamical Systems*, ed. M. M. Peixoto (New York: Academic Press, 1973), 195–233.

5. Hector J. Sussmann, "Catastrophe Theory," *Synthese* 31, no. 2 (1975): 229–271.

6. James Callahan, "Singularities and Plane Maps," *American Mathematical Monthly* 81, no. 3 (1974): 211–240. James Callahan, "Singularities and Plane Maps II: Sketching Catastrophes," *American Mathematical Monthly* 84, no. 10 (1977): 765–803.

7. Theodor Bröcker, *Differentiable Germs and Catastrophes*, trans. L. Lander (London: Cambridge University Press, 1975).

8. Thom, *Structural Stability and Morphogenesis*, 9. Emphasis in original.

9. Tim Poston and Ian Stewart, *Catastrophe Theory and Its Applications* (London: Pitman, 1978), 94.

10. Thom, *Structural Stability and Morphogenesis*, 15.

11. M. M. Peixoto, "Qualitative Theory of Differential Equations and Structural Stability," in *Differential Equations and Dynamical Systems*, eds. J. K. Hale and J. P. La Salle (New York: Academic Press, 1967), 469–481.

12. Zbigniew Nitecki, *Differentiable Dynamics* (Cambridge, MA: MIT Press, 1971).

13. David Berlinski, *On Systems Analysis* (Cambridge, MA: MIT Press, 1976).

14. Poston and Stewart, *Catastrophe Theory*, 67.

15. Tim Poston, "On Deducing the Presence of Catastrophes," *Mathématiques et Sciences Humaines* 64 (1978): 71–99, http://www.numdam.org/item/MSH_1978__64__71 _0.pdf.

16. Zeeman, *Catastrophe Theory*.

## 24. Mathematics and Its Applications

1. Charles Parsons, "Quine on the Philosophy of Mathematics," in Edwin Hahn and Paul Arthur Schilpp, *The Philosophy of W. V. O. Quine* (LaSalle, IL: Open Court Press, 1986), 382. Emphasis added.

2. Parson's phrase, the "instance of a structure" is not entirely happy. Predicates have instances; properties are exemplified; structures just sit there.

3. W. V. O. Quine, "Reply to C. Parsons," in Edwin Hahn and Paul Arthur Schilpp, *The Philosophy of W. V. O. Quine*, 398.

4. See Marceau Feldman, *Le Modèle Géometrique de la Physique* (Paris: Masson, 1992), 198–204, for interesting remarks.

5. By Euclidean geometry I mean any axiomatic version of geometry essentially equivalent to Hilbert's original system—the one offered in Chapter 6 of *Fundamentals of Mathematics*, eds. H. Behnke, F. Bachman, and H. Kunle (Cambridge, Massachusetts: MIT Press, 1983), for example.

6. Indeed, it is not clear at all that the surface of my desk is either a two- or a three-dimensional surface. If the curved sides of the top are counted as a part of the top of the desk, the surface is a three-dimensional manifold. What then of its *rectangular* shape? If the edges are excluded, where are the desk's *boundaries*?

7. In a well-known passage, Albert Einstein remarked that to the extent that the laws of mathematics are certain, they do not refer to reality; and to the extent that they refer to reality, they are not certain. I do not think Einstein right, but I wonder whether he appreciated the devastating consequences of his own argument? Albert Einstein, *Geometrie und Erfahrugen* (Berlin: Springer, 1921).

8. Quantum considerations, I would think, make it impossible to affirm any version of an Archimedean axiom for points on a *physical* line.

9. For very interesting if inconclusive remarks, see the roundtable discussion by a collection of Field medalists in C. Casacuberta and M. Castellet, eds., *Mathematical Research Today and Tomorrow* (New York: Springer-Verlag, 1991), 88–108, especially the comments of Alain Connes on page 95.

10. Defined as the ratio of two lengths, radians are in any case dimensionless units.

11. See, for example, Thomas Brody, *The Philosophy behind Physics* (New York: Springer Verlag, 1993), 56–58.

12. R. Larson, R. Hostetler, and B. Edwards, *Calculus* (Lexington, MA: D. C. Heath & Company, 1990), 122.

13. Hermann Weyl, *Symmetry* (Princeton, NJ: Princeton University Press, 1952), 45.

14. Curiously enough, this is a point that Weyl himself appreciates. See his discussion on pages 15–17 of his *The Classical Groups* (Princeton, NJ: Princeton University Press, 1946).

15. See S. Eilenberg, *Automata, Languages, and Machines*, vol. A (New York: Academic Press, 1974). From a philosophical point of view, interest in semigroups is considerable. A finite state automata constitutes the simplest model of a physical process. Associated to any finite state automata is its transition semigroup. Semigroups thus appear as the most basic algebraic objects by which change may abstractly be represented. Any process over a finite interval can, of course, be modeled by a finite-state automata; but physical *laws* require differential equations. Associated to differential equations are groups, *not* semigroups. This is a fact of some importance, and one that is largely mysterious.

16. This familiar argument has more content than might be supposed. It is, of course, a fact that quantitative measurements are approximate; physical predicates are thus *inexact*. For reasons that are anything but clear, quantitative measurements do not figure in mathematics; mathematical predicates are thus *exact*. It follows that mathematical theories typically are *unstable*. If a figure D just misses being a triangle, no truth strictly about triangles applies to D. Mathematical theories are *sensitive to their initial descriptions*. This is not typically true of physical theories. To complicate matters still further, I might observe that no mathematical theory is capable fully of expressing the conditions governing the application of its predicates. It is thus *not* a theorem of Euclidean geometry that the sum of the angles of a triangle is *precisely* 180 degrees; "precisely" is not a geometric

term. For interesting remarks, see Jacob T. Schwartz, "The Pernicious Influence of Mathematics on Science," in *Discrete Thoughts*, eds. Marc Kac, Gian-Carlo Rota, and Jacob T. Schwartz (Boston: Birkhäuser Boston, 1992), 19–25.

17. John Casti, *Alternate Realities* (New York: John Wiley, 1988), 22–25.

18. Wu-Ki Tung, *Group Theory in Physics* (Philadelphia: World Scientific, 1985), 2–5, whence my discussion.

19. For a more general account, see Hermann Boerner, *Representation of Groups* (New York: American Elsevier Publishing Company, 1970). For a (somewhat confusing) discussion of the role of groups in physics, see Victor Guillemin and Schlomo Sternberg, *Variations on a Theme by Kepler* (Providence, RI: American Mathematical Society, 1990).

20. Eugene Wigner, "The Unreasonable Effectiveness of Mathematics in the Natural Sciences," *Communications in Pure and Applied Mathematics* 13, no. 1 (February 1960): 1–14.

21. R. Courant and D. Hilbert, *Methods of Mathematical Physics*, vol. II (New York: John Wiley, 1972), 227. The notion of a solution to a differential equation is by no means free of difficulties. Consider a function $f(x) = Ax$, and consider, too, a tap of the sort that sends $f$ to $g(x) = Ax + mx$, where m is small. Do $f$ and $g$ represent two functions or only one?

22. Courant and Hilbert, *Methods of Mathematical Physics*, 227.

23. V. I. Arnold, *Ordinary Differential Equations* (Cambridge, MA: MIT Press, 1978), 97.

24. Michael Atiyah, *The Geometry and Physics of Knots* (New York: Cambridge University Press, 1990), 3.

25. Bas C. Van Fraassen, *The Scientific Image* (Oxford: The Clarendon Press, 1980), 64. See also W. Balzer, C. Moulines and J. Sneed, *An Architectonic for Science* (Dordrecht: D. Reidel, 1987) for a very detailed treatment of similar themes.

26. Dana Scott and Patrick Suppes, "Foundational Aspects of Theories of Measurement," *Journal of Symbolic Logic* 23, no. 2 (June 1958): 113–128, https://www.jstor.org/stable/2964389.

27. V. I. Arnold, *Mathematical Methods of Classical Mechanics* (New York: Springer Verlag, 1980), Chapter 1.

# Bibliographical Note

There follows a list of the essays from this volume that previously appeared elsewhere, along with their original place of publication. They are listed in the order they appear in this volume:

"Haunting History," "Misprints in the Book of Life," "A System of Belief," "The Evidence for Evolution," and "Partir c'est mourir un peu" (previously titled "Checking Out"), *Black Mischief: Language, Life, Logic, Luck,* 2nd ed. (New York: Harcourt Brace Jovanovich, 1988).

"Darwin and the Mathematicians," *Evolution News and Science Today,* November 7, 2009.

"Iterations of Immortality," *Harper's Magazine* (January 2000), 15–20.

"Responding to Stephen Fletcher's Views… on the RNA World Hypothesis," *Evolution News and Science Today,* January 15, 2010.

"The Activity of a Cell Is Like That of a Factory," *Evolution News and Science Today,* May 26, 2020.

"A Graduate Student Writes," *Evolution News and Science Today,* July 9, 2013.

"A One-Man Clade," *Evolution News and Science Today,* July 18, 2013.

"Good as Gould" and "Ovid in Exile," *Black Mischief: Language, Life, Logic, Luck,* 2nd ed. (New York: Harcourt Brace Jovanovich, 1988).

"A Natural History of Curiosity," *Inference* 1, no. 2 (March 2015).

"The Ineffable Higgs," *Evolution News and Science Today,* November 16, 2012.

"The Good Soldier," *Inference* 5, no. 2 (May 2020).

"Blind Ambition" and "Kolmogorov Complexity," *Black Mischief: Language, Life, Logic, Luck,* 2nd ed. (New York: Harcourt Brace Jovanovich, 1988).

"Review of Michael Ruse's *The Philosophy of Biology*," *Philosophy of Science* 41, no. 4 (December 1974).

"The Director's Cut," *Inference* 5, no. 1.

"*Catastrophe Theory and Its Applications*: A Critical Review," *Behavioral Science* 23, no. 4 (1978): 401–416.

"Mathematics and Its Applications," *Mathematics, Substance, and Surmise: Views on the Meaning and Ontology of Mathematics,* eds. Ernest Davis and Philip J. Davis (New York: Springer International Publishing Switzerland, 2015).

"Isaac Newton" (previously published as twin essays, "Newton's Version" and "Newton's Vision"), "The Ghost within the Machine," and "Life Itself," *Black Mischief: Language, Life, Logic, Luck,* 2nd ed. (New York: Harcourt Brace Jovanovich, 1988).

# INDEX